THE FREIGHT ONLY YEARBOOK

No 2

Michael Rhodes and Paul Shannon

SLP

Silver Link Publishing Ltd

First published in April 1991

A CIP catalogue record for this book is available
from the British Library.

ISBN 0 947971 59 9

Silver Link Publishing Ltd
The Trundle
Ringstead Road
Great Addington
Kettering
Northamptonshire NN14 4BW

Typeset by The TypeFoundry Limited, Northampton
Printed and bound in Great Britain by Woolnough Bookbinding Ltd, Irthlingborough, Northants

CONTENTS

INTRODUCTION

When we wrote the introduction to our first *Freight Only Yearbook*, we commented on the extensive changes which took place within Railfreight in 1989. Little did we realise that even greater changes were to take place in 1990. At an organisational level, the division of Railfreight into two distinct business units – Trainload Freight and Railfreight Distribution – became firmly established during the year. Trainload Freight comprises the bulk sub-sectors Coal, Construction, Metals and Petroleum, most of whose traffic can be moved in complete trainloads from siding to siding. Railfreight Distribution subsumes the Freightliner system and the Speedlink wagonload freight network, as well as certain trainload flows of chemicals, industrial minerals, edible products and motor vehicles.

The contrast between the fortunes of Trainload Freight and Railfreight Distribution was heightened in Railfreight's annual results published in spring 1990. Trainload Freight had a turnover of £516.7 million and achieved a healthy operating profit of £132.8 million. Railfreight Distribution had a turnover of £176.7 million but recorded an operating loss of £73.4 million. The comparison becomes all the more acute if Distribution's largely profitable Freightliner and trainload operations are taken out of the equation: Speedlink, it was revealed, was losing Railfreight £30 million on a turnover of £45 million. In former times the losses incurred by wagonload freight might well have been swallowed up in regional balance sheets; in the business-led Railfreight organisation of 1990, however, there was no possibility of cross-subsidy. So an exhaustive review of Speedlink was carried out, and the decision was taken in December to axe completely this sphere of Railfreight operations. The year 1990, therefore, was to be the last full year of wagonload freight on BR.

We have divided this volume into five main chapters, dealing with each Railfreight sub-sector in turn. The information contained in the introductory text to each chapter comes from a variety of sources. This year, however, we were especially grateful to Railfreight for granting us exclusive interviews with senior managers from each sub-sector. In addition to the five main chapters, we have included three extra sections, dealing with Class 60s, Class 50s, and departmental topics. Class 60 locomotives made appearances in many parts of the network but only a handful had entered revenue-earning service by the end of the year; we therefore felt that it would be more appropriate to group our Class 60 material together rather than try to distribute it among the sub-sectors. Whilst Class 50 locomotives have had no regular freight involvement for several years, they have attracted a considerable enthusiast following in the closing stages of their careers; we have therefore stretched a point and included a selection of engineers' trains – still freight even if not revenue-earning – hauled by these locomotives. Similarly, in our 'Departmental Interlude' we have included material which is not readily identifiable with any of the Railfreight sub-sectors, even though some of it – for example the movement of London Underground stock – belongs very clearly to the domain of revenue-earning freight.

We hope that the reader enjoys our review of Railfreight in 1990. Many of the scenes in this volume will be consigned to history within the next 12 months, with the abolition of Speedlink and the replacement of many older locomotives by Railfreight's new workhorse, the Class 60. Exciting times lie ahead for Railfreight Distribution as the opening of the Channel Tunnel draws nearer, and 1991 should see some additions to RfD's international portfolio in readiness for the eventual operation of through services. Finally we would like to offer our sincere thanks to the many railway employees, enthusiasts and photographers who have helped us to compile this book. We could not have done it on our own!

Paul Shannon, Bolton
Michael Rhodes, Newcastle-upon-Tyne
January 1991

Chapter 1

TRAINLOAD COAL

The privatisation of the Central Electricity Generating Board during 1990 presents Railfreight Coal with a new challenge to retain its preeminence in the supply of coal to domestic power stations. Railfreight Coal now has two contracts with National Power and Power Gen, both new companies on the alert for a competitive edge. The British coal industry faces a period of temporary stability, safe in the knowledge that quotas of power station coal have been fixed for three years from January 1990. In spite of this agreement there have been several new flows of imported coal during 1990, the majority rail-hauled from ports of entry.

IMPORT COAL

Imported coal has perhaps had the largest effect on the pattern of freight services provided by Railfreight Coal. The Gladstone Dock scheme in Liverpool has provided over 50 per cent of the coal required by Fiddlers Ferry power station and resulted in a major adjustment in train planning in the Warrington area. Spot coal, purchased from international stockpiles by Power Gen, has been imported through Liverpool, and Railfreight has been used for the last stage of the journey to Fiddlers Ferry. Initially four Class 20 locomotives were used to haul 46 HAA hoppers up from the dockside. September saw the introduction of pairs of Class 56 locomotives on the service, then in December the first Class 60 arrived on Merseyside and began trials on the Gladstone Dock services.

Precursors to the new Gladstone Dock scheme were the import of Chinese coal through Rothesay dock in 1988 and a similar import flow through Hull during 1989, bound for the Aire valley power stations. A new facility was established at Immingham in 1989 for British Steel's Scunthorpe plant, but this terminal may pale into insignificance if plans for a new dock at Immingham gain final parliamentary approval. This would clear the way for bulk imports to supply the Trent power stations at Cottam and West Burton. Other smaller import flows during 1990 have benefited from rail shipment. Coal arriving at Teesport has been carried to BS Scunthorpe, and coking coal has been imported at Hunterston and carried by rail to both Ravenscraig and Scunthorpe.

Railfreight Coal is eager to respond to changes in its customers' demands. Seizing new business opportunities obviously requires the injection of new investment, and key rail routes to receive attention during 1990 were those from potential coal ports. The plans for Immingham have led to both the retention of four tracks between Brocklesby and Barnetby and the decision to build a new curve to allow eastern access to Cottam power station (such access already exists at West Burton). Railfreight Coal has also started construction of a new train crew depot at Worksop. This will eventually cater for over 200 staff and provide the focal point for Railfreight Coal operations to Cottam and West Burton. Staff will transfer from the depots at Shirebrook and Barrow Hill to the new location which is closer to the purchaser of the coal and provides good access to both the local coalfield and potential import flows.

Change is also afoot in the Aire valley. Two schemes are under way to improve

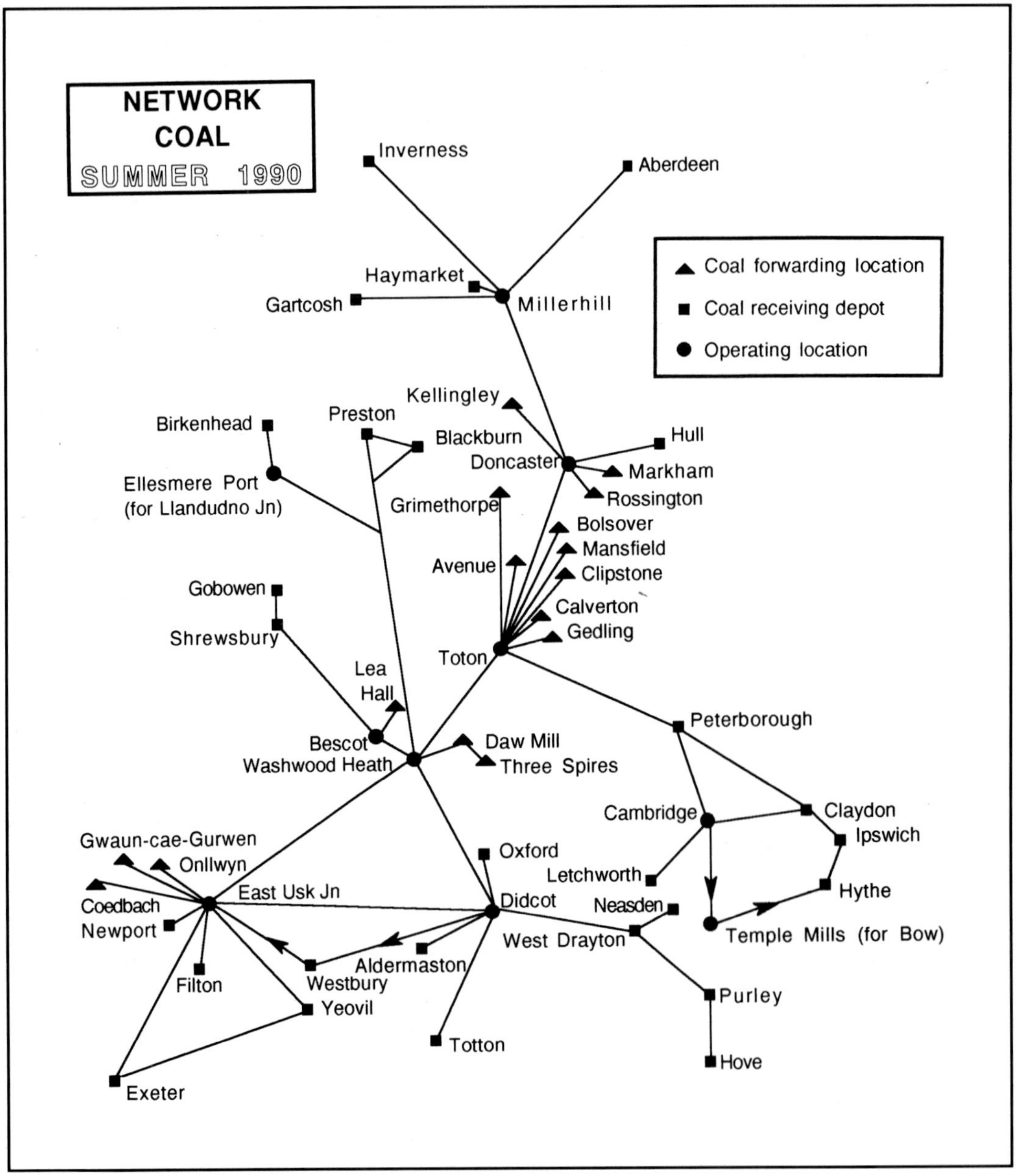

rail access to Drax, Eggborough and Ferrybridge power stations; access from Immingham, Scunthorpe and Askern is to be improved in order that imported coal might also be delivered by rail. Rail improvements should also simplify the routing of traffic from Selby drift mine to Drax power station. New curves are to be built at Heck and Hambleton Junction to allow traffic to reach Drax without having to traverse the busy route through Knottingley. Not only does this increase the number of journeys that can be made each shift but it also brings environmental benefits by reducing the number of MGR trains passing through the heavily populated Knottingley area. The Immingham dock scheme has already had a major impact on rail planning in Yorkshire. Associated British Ports have pursued this development with some vigour and there is speculation of further such coal import terminals at Milford Haven and Southampton.

NEW MINES

New rail routes are also planned or under construction for the new drift mines at Asfordby and Hawkhurst Moor. Asfordby mine should be producing 4 million tonnes of coal per annum by 1995, and with this in mind Railfreight Coal has announced its plans to connect the pit with the Old Dalby test track near Melton Mowbray and build a new west-facing chord onto the Peterborough to Leicester main line. Railfreight plans for Hawkhurst Moor are less advanced as British Coal has not yet obtained full planning permission for the new mine here; if it goes ahead, a new chord to Kenilworth is planned. A further drift mine at Margam has been announced by British Coal which is still negotiating with local miners regarding potential staffing arrangements at the new mine. Again Railfreight Coal plans to link any such mine to the rail network in order to supply both steel and generating industries.

SOUTH WALES AND THE NORTH EAST

Railfreight Coal has also been involved in several smaller schemes in the North East and South Wales. The Plenmellor terminal near Haltwhistle has been completed and a daily MGR service is due to start in early 1991. Similarly, 1990 has seen work begin in earnest on the Blyth dock scheme. This involves the re-establishment of a rail link to the River Blyth at Bates. Again, the new MGR train service is due to start in the spring of 1991. In South Wales there have been changes forced by the closure of pits at Blaenant and Bettws Drift. As in Scotland, the dwindling deep mining industry in South Wales has made way for an expanding network of opencast workings, the most recent of which is the new mine at Cwmbargoed. Indeed, it is hoped that this new site will provide up to eight trains a day for Aberthaw power station. This new opencast site has spurred investment in the branch from Ystrad Mynach, including the construction of a new passing loop at Tredomen and the installation of colour light signalling controlled from Ystrad Mynach South signal box.

Opencast coal is also mined around Onllwyn in South Wales but the rail service to the washery here has been severely disrupted by a strike which started on 23 May 1990. The strike continues at the time of writing – indeed, the local opencast sites had 200,000 tonnes of anthracite stockpiled by December 1990. The continuation of a rail service to the washery, owned by Ryan International, clearly depends on the outcome of one of the coal industry's longest and most bitter disputes. Stockpiled coal from another South Wales valley ceased during 1990, with the last trainload from Maesteg. The stockpile here took five years to remove after the closure of the pit!

SCOTLAND

The coal industry in Scotland has been reborn with the development of opencast mining in Ayrshire and Fife, and this has been parallelled by the recovery of Railfreight Coal services north of the border. The development of the Douglas basin and Chalmerston forms the backbone of coal operations in Ayrshire where Railfreight Coal is responsible for the delivery of 1 million tonnes of coal each year to Ayr docks whence it is exported to Northern Ireland. New markets for opencast coal mined at Westfield in Fife and Blindwells near Edinburgh provided Railfreight Coal with its most dramatic range of new services during 1990. Two daily trains of 36 HAAs from Westfield to Thornton had to be broken up into three trainloads of 24 HHAs over the Forth Bridge to Millerhill. From here two daily trains traversed the East Coast Main Line to Yorkshire power stations. These were joined by two further services originating from Blindwells – 4,000 tonnes of power station coal each day by rail.

PIT CLOSURES

The number of pit closures in 1990 was considerably less than in 1989 when 13 deep mines succumbed. Bettws Drift mine and Blaenant colliery both closed in South Wales whilst the railhead at Onllwyn has seen virtually no traffic in 1990 due to a strike. Perhaps the most significant closure of the year was at Abercwmboi. The British Coal phurnacite plant closed in March and stockpiled coal was finally cleared by August. Pits also closed at Shireoaks and Mansfield, making a total of four deep mine closures for the year.

CLASS 62

Railfreight Coal also announced the possibility of new motive power for some of its MGR coal services, not the delayed Class 60 but a new Class 62. As 1990 drew to a close the financial case for a new Type 5 locomotive was being finalised. The imminent need for half-life renewal of the Class 56 fleet raised the question of investment in a new 3,000 hp machine specifically suited to MGR operation. Such an engine would possess high adhesion, slow speed control and would tolerate permanent idling (necessary whilst standing at collieries or power stations). In addition, the locomotive would have to be immune to the ill effects of aggressive dust. Speculation that Railfreight Coal will look to North America for such a machine may well prove correct, particularly in light of the impressive performance of the Class 59s.

NEW WAGONS

In addition to new motive power, Railfreight has started to assess its wagon requirements for the 21st century. The HAA hopper has served the coal industry well for 25 years but is clearly a product of the 1960s. The problem of dust spillage from this open hopper is being addressed with wind tunnel studies on a new canopy which is claimed can reduce dust spillage to zero. There is also a rolling programme to rebody the HAA fleet, and this work will continue at the RFS works in Doncaster. The wagon is also being adapted for automatic vehicle identification (AVI); a pilot AVI system is currently being tested and refined in the Aire valley. In spite of all these improvements to the trusted HAA workhorse, these wagons cannot continue indefinitely.

Looking to the future, Railfreight hopes to employ a low track force bogie to improve efficiency in a new coal wagon. It would also be preferable if any new wagon could have a payload nearer 100 tonnes, in line with recent developments in the aggregates business. In order for such developments to be executed, careful planning to ensure terminal compatibility will be necessary. It is, however, clear that Railfreight has no intention of taking its success in the carriage of power station coal for granted. In addition to motive power and rolling-stock renewal, there are also plans to resignal key areas of the coal network. Both north Nottingham and Knottingley are high on the agenda for such modernisation.

COLLIERY WASTE, GYPSUM AND FLYASH

The carriage of colliery waste from Hatfield colliery to Glews Hollow near Goole provided a new train service in Yorkshire during 1990. A much bigger scheme to dispose of waste from Selby drift mine by dumping it near Normanton was inaugurated in September. British Coal's multi-million-pound Welbeck spoil terminal will receive about 1.25 million tonnes of spoil each year. This will necessitate between two and five daily trains from Selby to the site near Goose Hill Junction, each consisting of 47 HAA hoppers and carrying 1,500 tonnes of spoil. The colliery waste is being used to reclaim 870 acres of derelict land which will be upgraded for recreational use over the next 30 years.

Flyash traffic to Fletton, near Peterborough, will cease in early 1991. Alternative sites for dumping ash have been identified and these are close to the various power stations and therefore do not require rail transport. Whilst flyash traffic will decrease in 1991, there will be an exponential increase in the carriage of gypsum to Drax and Ratcliffe power stations as they adopt flu-gas desulphurisation to reduce acidic emissions. Railfreight should be the chosen method for this new traffic with daily trains to Ratcliffe and twice-daily services to Drax.

NETWORK COAL

The success of Trainload Coal is unfortunately not mirrored by its smaller partner, Network Coal. Railfreight emphasises that Network Coal has increased its market share in household and industrial coal distribution, but this market-place is shrinking fast. The 'Network' set up in 1987 has been vitiated during 1990. The withdrawal of trains to Purley, Hove and Totton brings an end to Network coal on the Southern Region, while further closures at Neasden, Ipswich, Cambridge and Peterborough have severely depleted wagonload coal services to East Anglia and Greater London. Network Coal has also pulled out of Oxford, and British Coal closures at Mansfield and Abercwmboi have further reduced the sources of supply to the network. Add to this the virtual cessation of HEA wagonload traffic from Deep Navigation, Onllwyn and Gwaun-cae-Gurwen, and it is not difficult to understand one senior British Rail official when he said that the network was 'dying by degrees'.

Coal is now collected from Coedbach in South Wales, three pits in the West Midlands, three in the East Midlands, and four in Yorkshire. Concentration of wagons at East Usk, Didcot, Washwood Heath, Toton and Doncaster enables supply to customers who remain in the north and west of the country.

It is quite remarkable how many things have changed during 1990. There have been closures and the loss of rail routes, but there have also been several areas of expansion and investment for Railfreight Coal. It seems certain that 1991 will continue to provide plenty of new services in what is a rapidly changing area of Railfreight operation.

Hove-Didcot: **November 1990 saw the end of coal deliveries to Hove coal depot which officially closed on 22 November. This picture shows what is believed to be the last train from Hove on 8 November 1990. Class 37 No 37213 passes Willesden South West with 6Z66, the 10.05 Hove to Didcot Network Coal service. *Brian Denton***

Ipswich: Class 37 No 37213 stands in Ipswich yard on 17 October 1990 with 6M03, the 03.40 from Temple Mills to Toton. The locomotive has just collected empty HEA hoppers from Claydon cement works. Its journey back to the Network coal sidings at Toton Old Bank will be via a stop at Letchworth (see below). This service, the last to East Anglia, looks vulnerable during early 1991 as several of the terminals it served have closed in the last 12 months. *Michael Collins*

Letchworth: Cambridge coal depot closed in early summer but the small depot at nearby Letchworth survived into 1991, served by the thrice-weekly Temple Mills to Toton which called as required. This necessitated a considerable diversion from the direct route via Ipswich and March. Class 31 No 31429 stands behind the disused goods shed at Letchworth on 23 March with three empty HEA hoppers and two loaded with coal from Calverton (see page 23). After propelling the loaded wagons into the discharge area (the hopper conveyor is visible above the end of the train) the locomotive will continue north as 6M03, the 03.25 Temple Mills to Toton. The use of a Class 31 was unusual; normally the train was hauled by a Network Coal Class 37. *Paul Shannon*

Foxton: Class 37 No 37214, owned by Railfreight Construction, passes bales of hay stacked near Foxton on 7 August 1990. The train is the 6M03 service from Temple Mills to Toton, now timed to depart from Temple Mills at 03.40 (see above). *Paul Shannon*

Radyr Yard: **At the start of 1990, Radyr yard in Cardiff was a busy Network Coal marshalling point. This view on 30 March shows Class 37 No 37239 easing out of the sidings with a lengthy 6C34 Radyr to Exmouth Junction service. In the background Class 08 No 08654 shunts empty wagons for the afternoon train to Coedbach. In July, Radyr yard closed to revenue-earning traffic and Network Coal moved into East Usk yard which had in the meantime been vacated by Speedlink (RfD). In January 1991 there were just four daily departures from East Usk with a similar number of arrivals.** ***Michael Rhodes***

Radyr-Yeovil-Exmouth Junction: **Class 37 No 37274 has just negotiated Dr Day's Bridge Junction, north of Bristol, on its way to Yeovil via Westbury. It heads 6C34, the 09.10 Radyr to Exmouth Junction. In July this train was transferred to East Usk yard with a departure time of 09.50 on MWFO.** ***Mike Goodfield***

NETWORK COAL : SURVIVORS

Bordesley: Another small coal depot to close almost unnoticed during 1990 was the LCP fuels sidings at Bordesley. On 18 March 1987, Class 08 No 08616 shunts HEA hoppers over the unloading bay. *Don Gatehouse*

Preston Deepdale: Whilst in many parts of southern England wagonload coal traffic has continued to decline, the two Lancashire depots at Blackburn King Street and Preston Deepdale remained relatively busy for much of the year. From October 1989 these depots were served from Washwood Heath instead of Healey Mills, with two out of four weekly trains calling at Blackburn as well as Preston. Class 37 No 37214 propels six empty HEAs away from Deepdale coal depot on 12 April 1990, before returning south as 6G32, the 10.22 Deepdale to Washwood Heath. The Class 37 illustrated here was soon reliveried in Construction colours (see page 10). *Paul Shannon*

Didcot-Westbury: Prior to November 1990, Didcot acted as an important Network Coal marshalling point where coal from Rufford to Westbury cement works was staged. On 2 June Class 56 No 56052 passes Didcot East with 6A04, the 09.50 SO Didcot TC to Westbury Cement Works train. From October, coal to Westbury travelled directly from Toton on a thrice-weekly service. It is anticipated that a similar twice-weekly service from the Midlands to Claydon cement works in East Anglia will be introduced during 1991 (see above). *Don Gatehouse*

'101 THINGS TO DO WITH REDUNDANT HEA HOPPERS'

Storage: On 18 July 1990 Class 56 No 56126 winds out of Tyne yard with the morning trip to Blyth Cambois. The first wagon in the train conveys lubrication oil whilst the second is for the collection of waste fuel at Cambois depot. The rest of the load is made up of redundant HEA hoppers which will be stored in Blyth West yard. The sidings here have considerable spare capacity since the closure of Blyth West staithes in December 1989. *Michael Rhodes*

TET testing: The decline in domestic coal traffic has left Railfreight Coal with a surplus of HEA hopper wagons. Some have found employment carrying other commodities such as rock salt and scrap metal, whilst others have provided a non-revenue-earning load for motive power trials. Most frequent beneficiaries of this surplus were the Class 60 locomotives which have invariably used HEA hoppers in their many test runs during 1990. Here on 30 May 1990, Class 87 No 87035 and Class 86/4 No 86425 pass Winwick Junction with empty HEAs whilst engaged in TET testing. *Paul Shannon*

SOUTH WALES : CLOSURES

Blaenant: **On 19 April 1990 British Coal announced the closure of Blaenant mine, the last in West Glamorgan. A new drift mine had opened in 1976 after the investment of £1.4 million to upgrade the original drift mine established in 1963. 590 men produced 500,000 tonnes of power station coal annually. Unfortunately during 1989 the mine had been plagued with geological faults which meant it failed to reach its 'break-even' point of 15,000 tonnes of coal a week. On 29 March Class 37 No 37702 has just arrived with the 6Z85 service from Barry. The locomotive will run round its train before loading at the discharge hopper in the background. The train will leave as a Class 7, restricted to 45 mph maximum speed for its journey to Aberthaw power station.** ***Michael Rhodes***

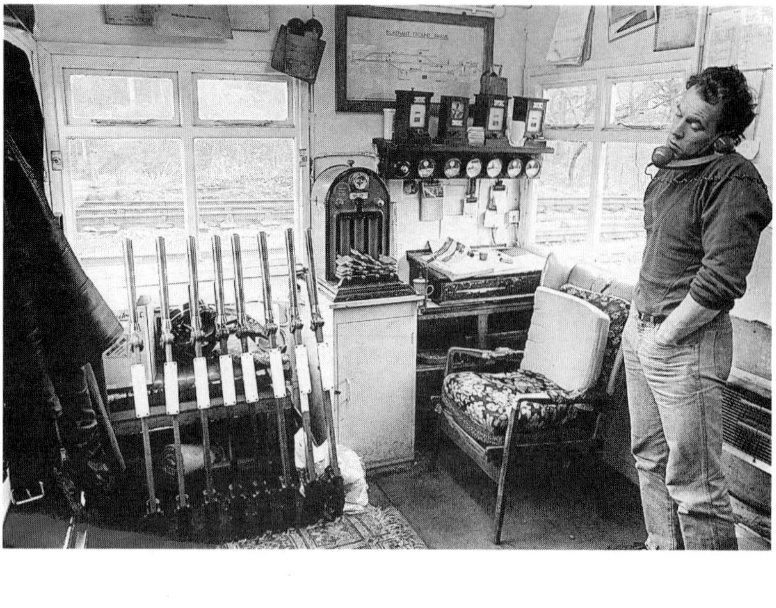

Blaenant signal cabin: **The lengthy branch from Swansea to Onllwyn is controlled by manual signal cabins at Neath and Blaenant. On 29 March the signalman checks the road is clear down to Swansea before releasing an MGR from the adjacent Blaenant pit. The future of both this cabin and the Onllwyn branch seemed far from secure at the end of 1990 because of a nine-month strike at Onllwyn washery and the closure of Blaenant mine.** ***Michael Rhodes***

Maesteg branch (Tondu): **The penultimate coal train from Maesteg to Newport Docks ran on 27 September 1990. The train is pictured here as it passes Tondu behind Class 37 No 37692. The surface coal at Maesteg has taken five years to clear, much of it travelling to Newport docks where it has been blended for Uskmouth power station. The line to Maesteg will survive in spite of the cessation of coal traffic, as Regional Railways have approved the plans to reintroduce passenger trains on the route. There will be stations at Pontyclun, Pencoed, Sarn, Tondu, Garth and Maesteg, all served by hourly trains to and from Cardiff.** ***Stuart Jolly***

Maesteg: **The washery buildings at St Johns (Maesteg) had been demolished by September 1990. On 10 September there were just 20,000 tonnes of coal to go before the job of clearing surface stocks was completed (see above). Class 37 No 37691 waits patiently for a solitary mechanical shovel to load its aging rake of MDV wagons.** ***Geoff Dowling***

Abercwmboi phurnacite plant: The ex-GWR signal cabin at Abercwmboi was painted grey in the early 1980s as British Rail capitulated to the hostile coal-ridden environment created by the adjacent coke works. The works were closed in March 1990 leading to cleaner air but considerably fewer freight services up the Cynon valley. On 4 August 1990 Class 37 No 37896 arrives at Abercwmboi with 6Z07, an extra service from Ellesmere Port. At this time there were only four trainloads of coke left at the site and the six HEA hoppers which stood loaded in the yard had sat motionless for two weeks awaiting orders for the stockpiled coal. Rail services to the plant ceased just four weeks later. *Michael Rhodes*

COLLIERIES UNDER THREAT

Deep Navigation: Deep mining is an unpredictable method of winning coal. All three pits in the Ystrad Mynach area have had temporary disruption of production during 1990 because of geological faults. Deep Navigation would seem to be the most vulnerable pit, producing just 600,000 tonnes of steam coal annually. Indeed, should the new \opencast developments at Cwmbargoed (just five miles north) be successful, it will be very difficult for deep-mined coal to compete for the power station market. On 6 August 1990 Class 37 No 37690 stands at Deep Navigation with 7C87, the lunchtime train to Aberthaw. *Michael Rhodes*

Penallta: Penallta colliery was opened in 1905 and lies one mile north of Ystrad Mynach. The mine lies at the end of the old freight-only branch to Penrhiwfelin goods depot. On 28 March 1990 Class 37 No 37896 reverses into the pit which is seen on the right of this view taken from the end of the branch line. After loading it will depart as 7C86, the afternoon MGR to Aberthaw. 1990 has been a good year for Penallta which normally produces 500,000 tonnes of power station coal but has exceeded this amount, necessitating many extra MGR services. *Michael Rhodes*

NEW DEVELOPMENTS

Swansea Burrows yard (1): **The yard at Swansea Burrows handles mainly coal traffic for the port and Railfreight Distribution wagons from the nearby Ford plant and Gower chemicals. Both the yard and the neighbouring docks retained much of the rambling rail network built up in the first part of the century when Swansea was a busy port. In October 1990 Railfreight Coal rationalised the rail layout in the area, involving the reduction of Swansea yard to just ten sidings and the abolition of the ex-Great Western signal cabin at Swansea Burrows. Access to the dock area was also simplified, signalling for the eastern approaches of the yard was transferred to Port Talbot panel box, and the turnouts within the yard and dock area were converted to hand-points. This sadly meant the disappearance of the vintage Great Western lower quadrant signals seen in this view of the yard taken on 7 August 1990. Class 08 No 08993 (with a cut-down cab for operation to Cwmmawr) awaits its next duty in the yard, which is filled with a variety of traffic. On the left is the evening departure for Dagenham made up of a mixture of tank wagons and car carriers. Both the new MEA mineral wagons and vintage MDV vehicles are still in use in the centre of the yard whilst timber, petroleum, chemical and automotive wagons fill the right-hand side of this view.** ***Michael Rhodes***

Swansea Burrows yard (2): **The fine array of signal levers at Burrows sidings now lies out of use. The box contained one of the largest ex-GWR frames to survive into the 1990s.** ***Michael Rhodes***

Ystrad Mynach: The increase in opencast mining in the hills to the north-east of Merthyr may mean as many as eight new MGR services each day between Cwmbargoed and Aberthaw. To cater for this anticipated increase in traffic Railfreight Coal has invested in a new signalling panel which will be installed alongside the ex-Rhymney Railway frame at Ystrad Mynach seen being worked by signalman Watkins on 28 March 1990. Work started on the new panel in October and signalling alterations should be complete by July 1991. *Michael Rhodes*

New wagons – MEA/MAB: A new fleet of air-braked mineral wagons has been constructed during 1990. Based on the HEA chassis, they are designed for loading and unloading by mechanical scoops and have worked predominantly in South Wales. The MEA wagon has been constructed by RFS Industries in Doncaster whilst the MAB variety shown here is BR-built. The main routes to benefit from these new vehicles are into Swansea, Cardiff and Newport Docks. *Michael Rhodes*

MEA prototype: The forerunner of the new MEA wagon was the PMA which was originally built in the 1960s by Standard Wagon at Reddish and Heywood. An example of this vehicle was tried out on coke and household coal traffic in South Wales as long ago as 1986. The wagon is seen here at the front of 6C94, the afternoon Coedbach to Severn Tunnel Junction service. The train is seen leaving Kidwelly behind Class 37 No 37181 on 3 July 1986. *Michael Rhodes*

Trelewis Drift: The mine at Trelewis closed in 1989 but trains continued to run during 1990 in order to move a 200,000-tonne stockpile of surface coal. On 6 August 1990 Class 37 No 37696 stands under the derelict screen at Trelewis with a rake of newly introduced MEA wagons. It will draw forward under the rapid loading bunker at Taff Merthyr colliery which is being used for Trelewis coal, then will depart as the 7B24 service to Newport Docks where the coal will be blended for Uskmouth power station. The introduction of MEA mineral wagons for coal flows to Newport and Swansea Docks will allow the withdrawal of the older MDV-type wagons still used on these trains. *Michael Rhodes*

Rufford to Newport Docks: Th new MEA wagons were not confined to South Wales in 19 They were also utilised on ne flows of coal from the East Midlands to the blending sites both Cardiff and Newport Doc On 6 June 1990 Class 37 No 3 passes St Andrews in Birming with a Newport Alexandra Doc Rufford coal empties. *Geoff Dowling*

Railfreight Coal – Barry: The Railfreight Coal-sponsored depot at Barry introduced an attractive new logo during 1990. This is to be applied to all stock allocated at the depot. Amongst the first vehicles to carry the galleon emblem were the new MEA wagons. *Michael Rhodes*

SOUTH WALES : SURVIVORS

Tower: **As 1990 drew to an end, the colliery at Maerdy finally closed, bringing to an end coal mining in the world-famous Rhondda valley. Since 1986 coal from Maerdy had been brought to the surface at Tower colliery, near Aberdare, which now provides the only freight traffic north of Radyr. There are once- or twice-daily MGR services to Aberthaw power station. The pace of change in the South Wales coalfield is dramatic; it was only ten years ago that coal trains would queue five deep between Taffs Well and Radyr, but the 17 collieries which fed into Radyr yard in 1977 have now been reduced to just one! On 8 August 1990 the daily empty MGR to Tower colliery winds past Cwmbach behind Class 37 No 37897. The trees in the background shroud the derelict coke works at Abercwmboi.** ***Michael Rhodes***

MDV Mineral wagons: **Thirty-year-old MDV No 310810 is the last vehicle in a Swansea to Coedbach coal service which was photographed in Burrows sidings during August 1990. These aged workhorses will all have been withdrawn by the end of 1990.** ***Michael Rhodes***

THE MIDLANDS

Shirebrook: **1990 was the last year for Shirebrook train crew depot. The staff are due to transfer to a newly constructed depot at Worksop early in 1991, but the locomotive servicing facilities are to be retained in the short term at least and engines will use the depot for minor repairs and refuelling. On 3 May 1990 the depot is host to a range of Railfreight Coal motive power including, from left to right: Class 58 Nos 58026 and 58041, in the shed Class 58 No 58034 and Class 20 Nos 20210 and 20214, and finally on the right Class 58 No 58029.** ***Michael Rhodes***

Calverton: **The last surviving pit served by the once busy ex-MR branch from Nottingham to Annesley, Calverton provides one or two trains of power station coal daily for Ratcliffe power station. Also there are occasional HEA wagonloads of household coal which connect with the national network at Toton yard. On 3 May 1990 Class 20 Nos 20187 and 20052 ease through the old exchange sidings with 6T62, the afternoon trip from Toton. This will be loaded by a rapid loading bunker behind the photographer,depart about an hour later for Ratcliffe.** ***Michael Rhodes***

Clipstone South Junction: **Clipstone Junction signal cabin lies in the heart of the Nottinghamshire coalfield and controls a triangular set of junctions permitting access to mines in the Mansfield area. The through route south to Toton closed in the early 1980s and by 1986 the sidings at Mansfield were also redundant; however, the mines and coal-stacking sites at Mansfield, Rufford, Clipstone and Bilsthorpe continue to handle rail traffic. Trains to and from Clipstone and Rufford are predominantly of household coal, as in this view taken on 3 May 1990. Class 20 Nos 20053 and 20094 have had to run from Shirebrook to Thoresby Colliery Junction and then run round their train, the 7T62 from Toton, because of a points failure at Clipstone West Junction. They are seen here passing Clipstone South Junction with a rake of HEA hoppers returning from Westbury cement works to Clipstone colliery.** ***Michael Rhodes***

Gedling: **One of the smaller pits in Nottinghamshire is at Gedling. The 600,000-tonne annual output is mainly for the domestic market and there are therefore no MGR services from the mine. On 3 May 1990 Class 20 Nos 20053 and 20094 have just arrived in the colliery exchange sidings with the 6T65 morning trip from Toton. The load of HEA and HAA hoppers will be loaded ready for the next morning whilst the locomotives will return to Toton with eight HAA hoppers loaded with coal for Aldermaston and Padworth.** ***Michael Rhodes***

Barrow Hill yard: The sidings at Barrow Hill were once extensive with the main yard on the south of the main line containing nearly 20 roads. Over the last ten years the yard to the north of the main line has disappeared and the up yard has been reduced to a handful of recess sidings for MGR train sets; very little coal traffic travels in wagonloads and there is therefore no need for the once ubiquitous marshalling sidings. The yard pilot at Barrow Hill was withdrawn in 1987 and it is likely that all MGR staging will be transferred to Seymour yard during 1991. On 4 July 1990 Class 58 No 58039 storms out of Barrow Hill yard with the lunchtime MGR to Didcot power station. At the time of writing nearly all the coal for Didcot was being mined at the nearby Markham colliery leading to as many as eight daily MGR services from Barrow Hill to Didcot. These trains are likely to be manned by Toton crews from January 1991, as the locomotive depot at Barrow Hill is to close. *Michael Rhodes*

Barrow Hill depot: Barrow Hill roundhouse, opened in 1870 by the Midland Railway, is the last operational roundhouse in Britain, and for the last 20 years it has acted as a stabling point for locomotives involved in local coal traffic. The concentration of Railfreight Coal staff at Worksop planned for January 1991 will lead to its closure. On 4 July 1990, Class 58 No 58044 stands solitary in the rain at Barrow Hill. The many puddles on the turntable and surrounding floor are testament to the poor state of the glass roof. *Michael Rhodes*

Worksop: Train crew for the heavy MGR traffic to Cottam and West Burton will be based at Worksop from January 1991. On 13 November 1990 Class 56 No 56022 winds out of the up yard at Worksop with a Harworth to West Burton MGR service. The up yard has 20 sidings to cater for any potential increase in trainset stabling. The locomotive stabling point and train crew facility will be on the other side of the main line, adjacent to the wagon repair shops. Three sidings have already been identified as suitable for stabling up to 20 Type 5 locomotives. *Michael Rhodes*

Coventry-Tring: Merry-go-round coal trains continued to venture down the West Coast Main Line during 1990, travelling as far as Pitstone cement works near Tring, and providing the sight of a Class 58 at such unlikely places as Bletchley and Leighton Buzzard. On 1 August, the hottest day of the year when the temperature exceeded 100°F in Cheltenham, Class 58 No 58017 passes Leighton Buzzard with 7R19, the 10.13 Three Spires Junction to Tring service. Pitstone cement works, incidentally, ceased forwarding cement by rail during 1990. *Paul Shannon*

THE NORTH WEST

Gladstone Dock (1): **Imported coal was transported by rail from Gladstone Dock to Fiddlers Ferry power station during 1990. The traffic started in 1989 but continued during 1990 on an increased scale, the actual number of trains being dependant on the arrival of ships in Liverpool Docks. For the first six months of 1990, trains were hauled by two Class 20s between Edge Hill and Fiddlers Ferry and by four locomotives between Edge Hill and Gladstone Dock. In summer 1990 trails began using two Class 56 locomotives for the approach to the docks, and by the end of the year the first Class 60 had arrived on Merseyside to begin work on this most arduous diagram. On 26 May Class 20 Nos 20168, 20128, 20187 and 20052 are seen approaching Gladstone Dock with the early morning empties from Edge Hill.** ***Paul Shannon***

Gladstone Dock (2): **The first appearance of a pair of Class 56 locomotives on a revenue-earning train in Liverpool Docks took place on Saturday 28 July. Class 56 Nos 56023 and 56014, both normally Yorkshire machines, hauled 45 empty HAAs from Edge Hill to Gladstone Dock and returned an hour and a half later with their train fully loaded. The inward working is illustrated here running alongside Regent Road, the main dockland thoroughfare.** ***Paul Shannon***

Gladstone Dock-Fiddlers Ferry: **Prior to the introduction of MGR services from Liverpool to Fiddlers Ferry, the freight-only line past Widnes West Deviation Sidings was very quiet; just two or three Speedlink trains used the line each day. On 22 February 1990 Class 20 Nos 20045 and 20159 pass the sidings with 7R67, the 09.45 Gladstone Dock to Fiddlers Ferry train. The catenary of the electrified main line between Liverpool and Crewe can be seen in the background.** ***Paul Shannon***

Maryport-Rugeley: **A daily trainload of opencast coal continued to run from Maryport to Rugeley power station during 1990, often with an extra working on Saturday. As well as being the only coal train to traverse the Cumbrian Coast line, this train is unique within Railfreight Coal in that it is diagrammed for Class 47 haulage. Since Coal does not own any Class 47s, the normal arrangement is to borrow a locomotive from the Chemicals and Industrial Minerals Division of Railfreight Distribution. On 24 July 1990 7K31, the 18.30 Maryport to Crewe Basford Hall, is seen winding past Parton.** ***Paul Shannon***

THE NORTH EAST

Wardley opencast disposal point: On 6 November 1990 Class 56 No 56130 was named after the Wardley terminal. The disposal point opened over a year ago and caters for up to five trains each day. Most of the opencast coal from the site is transported to Tyne Dock. *Ian Carr*

Hatfield colliery: In 1990 Railfreight has benefited from two contracts to carry colliery waste to dumping grounds. The largest scheme is based on a new terminal at Normanton (see above). A smaller scheme which involved trains from Hatfield colliery to Glews Hollow near Goole is illustrated here. Class 31 No 31286 stands at the head of a rake of PTA and POA wagons being loaded at Hatfield colliery on 4 April 1990. *Les Nixon*

Pespool tip (1): The closure and rationalisation of the deep mining industry has led to the closure of many fascinating colliery railways. One survivor is the British Coal line to Pespool tip, which occupies the northern end of the old through route from Murton colliery to Hartlepool. Just over one mile of track still survives to connect the Pespool tip to Murton colliery, and several trains each shift carry colliery spoil to the tip in a variety of British Coal hoppers. On 14 June 1990 Barclay No 594 *Big Momma* is about to leave the sidings at Pespool with empty wagons for Murton. The wagon tipper can be seen in the background. *Michael Rhodes*

Dawdon colliery: British Coal announced the closure of Dawdon colliery during summer 1990, although the mine will not close completely until April 1991. During 1990 it produced just under 1 million tonnes of coal, much of which travelled south to Selby where it was used to 'sweeten' the output from the Selby drift mine. On 14 June 1990 Class 56 No 56115 starts to reverse under the rapid loader at Dawdon with the 6J09 turn from Sunderland Docks. On this occasion the coal is bound for Tyne Dock from where it will be shipped to the Thames estuary power stations. *Michael Rhodes*

Pespool tip (2): The unloading facilities at Pespool utilise an old-fashioned wagon tippler. Such equipment has been rendered almost obsolete since the introduction of bottom discharge hoppers like the HAA. *Michael Rhodes*

Sunderland-Selby: **The need to 'sweeten' coal mined at Selby is the sole reason for up to ten MGR services each day from the North East coalfield. Trains run mainly from the Durham coalfield although diagrams exist for them to work from both Wardley and the Blyth and Tyne area. On 22 May 1990 Class 56 No 56134 thunders past Long Lane signal cabin on the freight-only line from Eaglescliffe to Northallerton. Manual signal cabins survive at Picton and Long Lane on this route.** ***Michael Rhodes***

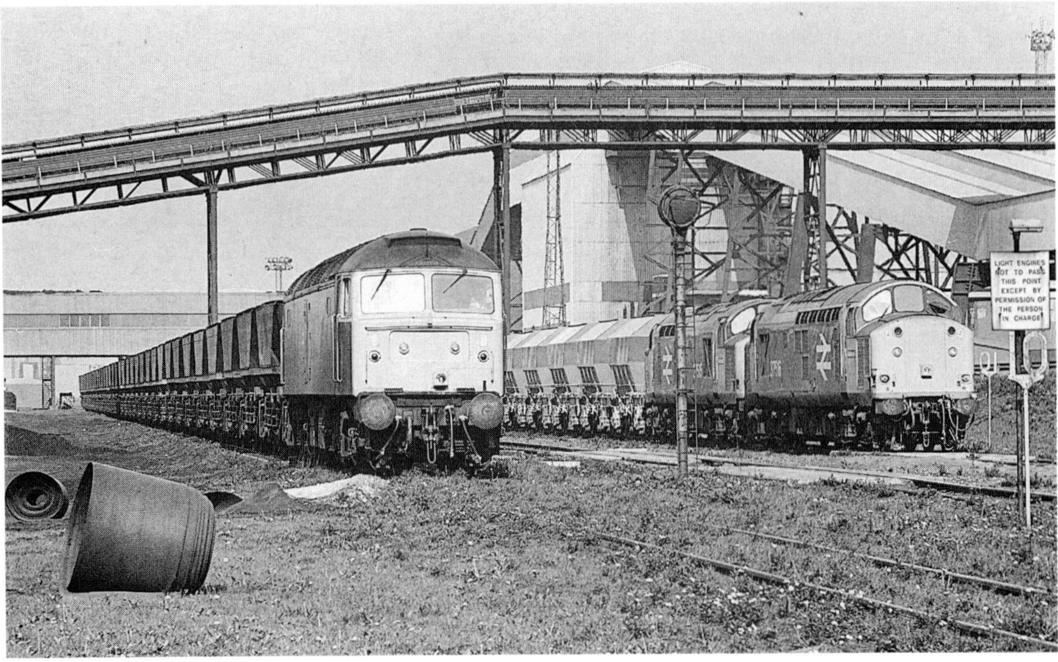

Teesport: **During the spring of 1990 a short-term contract to move imported coal to BS Scunthorpe brought extra rail traffic to the Redcar Mineral Terminal. On 22 April Class 47 No 47304 stands with an MGR to Scunthorpe which will be loaded with import coal, whilst to the right Class 37 Nos 37516 and 37515 await departure with 6N53, the daily train from Redcar to the British Steel-owned quarry at Redmire.** ***Michael Rhodes***

SCOTLAND

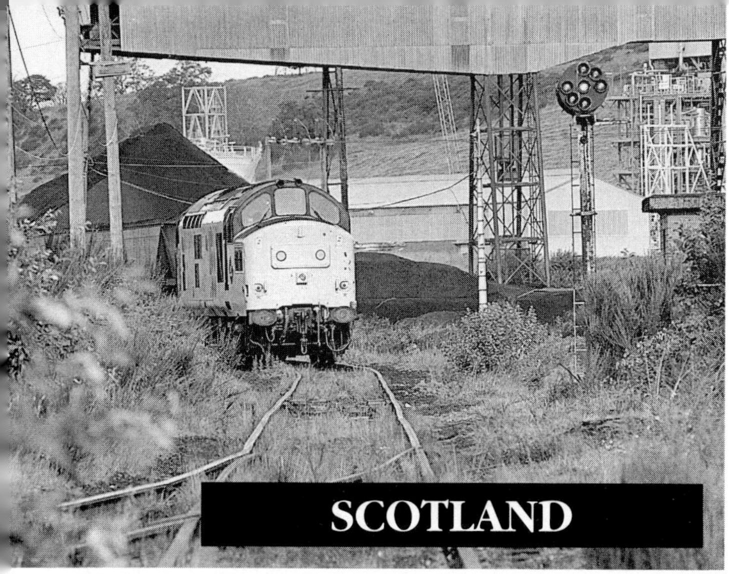

Westfield: The contract to move two trainloads of coal a day from Westfield opencast site in Fife had a profound effect on the freight timetable in Scotland, with two daily trips being reintroduced on the Westfield branch near Thornton. At Thornton these two trainloads of 36 HAA hoppers were remarshalled into three 24 HAA trains to traverse the Forth Bridge on their way to Millerhill, where they were reformed into two 36 HAA trains. On 17 August 1990 Class 37 No 37380 stands at Westfield with 7G01, the morning trip to Thornton. *Michael Rhodes*

Thornton-Millerhill: 6Z76, the 14.49 Thornton to Millerhill MGR service, is seen approaching the Forth Bridge at Inverkeithing. Class 56 No 56100 heads 24 HAA hoppers which are the third train of the day to Millerhill. *Paul Shannon*

Westfield branch: One of the last vacuum-braked workings in the country departs from Thornton yard on 27 August 1990. Class 37 No 37262 *Douneray* heads 7G01, the Methil power station to Westfield. The train consists of 28 MSV wagons and will be loaded with slurry at Westfield. The power station at Methil is specifically designed to burn low-grade coal slurry. *Paul Shannon*

Chalmerston branch: **The Chalmerston branch is rarely photographed because the services which run on weekdays operate overnight. On Saturday, however, there are two morning trips down the branch from Ayr. On 1 September 1990 Class 37 Nos 37694 and 37693 head north along the branch with 32 HAA hoppers loaded with opencast coal for washing at Killoch.** ***Paul Shannon***

Ayrshire-Margam: **The expansion of opencasting in Scotland has opened many new markets for the cost effective coal being produced. One short-term flow from Ayrshire was to BS Margam, and Railfreight was again the chosen method for transporting the coal, as on the east coast between Westfield/Blindwells and Yorkshire. On 15 June 1990 Class 37 No 37798 passes Bordesley Junction in Birmingham with 6Z10, an empty coal train from Margam to Carlisle. The 28 HDA wagons will be taken forward to Carlisle by a Scottish Railfreight Coal locomotive, whilst the Class 37 illustrated here will return south with a new load of coal.** ***Don Gatehouse***

Hunterston Low Level: **The branch to Hunterston Low Level is rarely used, but in 1990 there was a flow of imported coke for BS Scunthorpe which entered the country at Hunterston and utilised Railfreight for the southward journey to Scunthorpe. On 24 April 1990 Class 37 Nos 37376 and 37051 wind off the jetty at Hunterston with 6Z69, the 13.35 to Scunthorpe CHP.** ***Bob Avery***

Roughcastle: **One of Scotland's smaller opencast sites is at Roughcastle, near Falkirk. Class 56 No 56073 is seen on 12 October 1990 as it runs round an extra MGR service from Roughcastle to Millerhill.** ***Tom Noble***

Chapter 2

TRAINLOAD CONSTRUCTION

After six successive years of rising tonnages, peaking at around 16 million tonnes per annum, Trainload Construction was faced with a sharp fall in aggregates volume during 1990 as a result of the downturn in the national economy. Trainload Construction was the worst hit of all the Railfreight sub-sectors because 75 per cent of its traffic in recent years has been aggregates for South East England, and this is precisely the area which saw the biggest decline in construction activity, following the collapse of property prices during 1989.

Despite this rather gloomy background, Trainload Construction was able to record a number of successes during the year. Backed by a Section 8 Grant worth more than £1.5 million, an entirely new aggregates terminal was opened by Bardon (London) Ltd at Angerstein Wharf, South London. The new terminal was connected to the BR network in June 1990 and trains started running in the following month, well in advance of the official contract which was due to start at the end of the year. The terminal is fully equipped with a hopper discharge pit, extensive storage facilities and a wagon weighbridge. The traffic handled there comprises roughly equal quantities of limestone from the Mendips, loaded on to rail wagons at Foster Yeoman's Merehead quarry, and granite from Bardon Hill quarry in Leicestershire. The location at Angerstein Wharf was chosen to complement Bardon's other main receiving terminal at Thorney Mill on the western side of London. The Thorney Mill terminal was rebuilt to handle modern wagons in 1986, and here too there are regular flows from both Merehead and Bardon Hill.

Another successful acquisition by Trainload Construction was a new flow of limestone from the Steetley quarry at Whitwell, Derbyshire, to a discharge terminal at Witton, near Birmingham. This was significant because Steetley was previously the only large quarrying concern with no regular rail involvement. The new flow made use of existing sidings at both ends of the route. The terminal at Whitwell had previously been used to dispatch wagonload traffic to mainland Europe, whilst the sidings at Witton were the principal destination for trainloads of limestone from Caldon Low until that flow ceased in 1989. The new traffic from Whitwell had started to run on a trial basis to Wednesbury in the latter part of 1989, but was transferred to Witton in early 1990. A timetable for three trains a week was drawn up, although tonnages increased during the summer to the extent that sometimes four or even five trains were necessary. Traction was taken from the pool of Trainload Construction Class 56 locomotives based at Toton, and the wagons chosen for the new service were flat-bottomed POAs owned by the wagon-leasing company Tiger Rail. There was no necessity to provide hopper wagons because the limestone is loaded and emptied at each end of the route by mechanical grab. Given the success of the Whitwell-Witton operation it might be expected that Steetley would extend its use of rail to other parts of the country during 1991, especially if the construction industry regains some of the momentum lost in 1990.

The Redland self-discharge trains continued to work from Mountsorrel to Broxbourne, Langley Junction, Chesterton Junction, Banbury and Chelmsford. An additional destination for 1990 was Cricklewood, where space was made available in the former down sorting sidings to act as a distribution point for North London. The flexibility of the self-discharge concept was also appreciated by BR's civil engineers, who regularly hired a rake of wagons at weekends for track maintenance and relaying work on the Southern and Western Regions. To cater for the increase in traffic, Redland commissioned a total of 40 extra self-discharge wagons during the year, bringing the total number in use to 160. Redland also established a small terminal at Slough during 1990, but this was to be served by conventional trains from Whatley quarry, unloaded at Slough by mechanical grab. An order was placed with Powell Duffryn Standard for a rake of self-discharge vehicles for use in France, and this order was duly fulfilled in 1990.

Two new aggregates flows which started in late 1989 but became more regular in 1990 involved the reopening of the former coal concentration depot sidings at Limbury Road, Luton, where coal traffic had ceased in 1986. The new flows comprised gravel from Marks Tey and granite from Tarmac's recently commissioned terminal at Stud Farm, Leicestershire. Careful scheduling enabled both flows to be covered by the same traction and rolling-stock that was used on the long-established flow of gravel from Marks Tey to Mile End. Loaded trains from Marks Tey to Luton took a surprising route: instead of running via the North London line they were routed via Ipswich, Ely, Peterborough and Leicester.

Limestone from the Mendips continued to make Westbury one of the busiest freight centres on the Western Region. Foster Yeoman's five Class 59 locomotives still performed impressively in their fourth year of service, and broke new ground during the year when they started appearing on trains to Woking via Salisbury and the former London & South Western main line. Rival firm ARC's own fleet of four Class 59s arrived at Newport docks in October, and all four were ready to enter traffic in late November after completing tests at Derby Railway Technical Centre. Major changes to ARC's train plan would result from the introduction of the Class 59s, with fewer but longer trains on several routes. In December, for example, double-length trains started running from Whatley to Acton, where portions were divided for Allington and Chislehurst. Previously Allington had been served independently from Whatley. The receiving terminal at Chislehurst had traditionally been paired with Ripple Lane, with between 11 and 14 wagons going to each destination. ARC is opening a new plant at Dagenham to replace Ripple Lane and this flow will also feature in mega-train operation from Whatley. The relatively self-contained flow from Whatley to Theale was another one where the benefits of Class 59 haulage soon became apparent. Here train lengths were almost doubled, and frequency correspondingly reduced.

ARC was also able to boast a major development in its rolling-stock fleet during 1990. The first of an order of 145 hopper wagons incorporating a new design of low track force bogie took to the rails during the summer. Coded JHA and with a gross laden weight of 102 tonnes, the new wagons will cause significantly less track wear than the conventional bogie hoppers that ARC has used in the past. The contract for the new wagons was divided between two companies, with Powell Duffryn Standard building 123 of them and Procor the remaining 22. Other rolling-stock developments during the year included the delivery of 13 JHA hopper wagons built by W. H. Davis for Marcon at the beginning of the year, and the ordering of 25 90-tonne hopper wagons from RFS Industries for use on RMC stone trains between Peak Forest and Ely.

Three roadstone flows came to an end in 1990. These were Mountsorrel to Aylesbury, Tunstead to Widnes and Tunstead to Pendleton. The end was not necessarily permanent, however, since the wagons used on the Tunstead services were being stored pending possible re-entry to service, and it was possible that Aylesbury might become another destination for Redland's self-discharge train at some point in the future.

The movement of construction materials for the Channel Tunnel reached its peak during 1990. Railfreight has played a significant role in moving cement, concrete segments, crushed rock, minestone and steel for use in the project, but with the bulk of the business now completed, 1991 will see a reduction in tonnages and 1992 is expected to see only a small amount of residual traffic. Celebrations were the order of the day on 8 November when Railfreight operated the 1,000th train of concrete lining segments from Grain to Shakespeare Cliff. Looking ahead to the more permanent traffic which will eventually use the tunnel, the first trackwork was installed at Dollands Moor during 1990, providing access to what will become the principal interchange and customs point on the English side of the Channel.

Household and industrial waste was one area of Trainload Construction's operations which did not suffer as a result of the economic downturn. Long-term flows continued to operate from three refuse transfer stations in London, three in Manchester and three in Avon, plus the new flow from Edinburgh which commenced at the end of 1989. One feature of 1989 was the temporary flows of industrial waste from two development sites in London, at Barking and Willesden. In 1990 Railfreight again demonstrated its flexibility when it acquired a third short-term contract, this time to remove waste materials from Chatham Dockyard. The traffic from Chatham commenced operation in late October, and was expected to last for 12 months. Its destination was the landfill site at Forders Sidings, Stewartby, as was the case with the other two flows. The Chatham operation was more complex, however, in that the Dockyard branch is subject to a severe weight restriction and the journey to Stewartby involves rather more pathing constraints on the busy lines across London. The timetable was conceived in two parts, using Hoo Junction yard as a staging point. Up to nine trip workings are scheduled to run each day from Chatham to Hoo Junction, and these are then combined to form up to three trunk trains from Hoo Junction to Stewartby. Some trains are routed via the Midland route and others via the West Coast Main Line.

Cement carryings were roughly stable in 1990, having declined sharply in 1987 and 1988 after the ending of the cement industry's common pricing agreement. Basically it is now uneconomic for cement firms to transport their product over long distances, yet it was precisely the longer-distance flows where rail had its largest market share. Despite this, one cement terminal was reopened during the year, this being the Castle terminal at Birmingham Curzon Street. The traffic here comes from Padeswood, or Penyffordd in railway parlance. The Castle cement works at Tring ceased forwarding cement by rail in May 1990, the last regular flow having been a weekly train to King's Cross. In future King's Cross would receive all its supplies by rail from Ketton, whilst Tring would concentrate on more immediate markets in the South East – to be served by road.

Trainload Construction expects to acquire a fleet of 31 Class 60 locomotives in due course. According to the sub-sector's plans in December 1990, this figure will include 12 for Leicester, five for Hither Green, five for Westbury and seven for Buxton. The main allocation base is likely to be Toton, with the other depots being used for stabling, fuelling and minor examinations. In some cases Class 60 haulage will permit an increase in train length. Amongst the first flows to benefit from the new traction are expected to be those from Mountsorrel to the South East and East Anglia. At Westbury the Class 60s will work alongside the privately-owned Class 59s and displace the reduced number of Class 56s which are retained there at the time of writing. In advance of the arrival of the Class 60s, Trainload Construction has been actively pursuing a policy of traction replacement on key services in the London area. In the main this meant replacing a pair of Class 31s or a pair of Class 33s with a single Class 56 locomotive. By December 1990 Class 56 haulage was the norm on all the domestic refuse trains, on the Marks Tey-Mile End and Fen Drayton-King's Cross sand trains, and on a number of Southern Region services. The only train to retain double-headed Class 31s as traction was the cement working from Ketton to King's Cross.

ARC'S NEW LOCOMOTIVES

Newport Docks: **Class 59/1 No 59104 has its covers removed by ARC, General Motors and BR inspection staff at Newport Docks on 20 October 1990. After being passed for movement on BR metals, this locomotive and the other three Class 59/1s went to Whatley quarry where basic equipment commissioning was carried out. *Colin J. Marsden***

Saxby: **ARC's four Class 59/1s were taken to the Railway Technical Centre at Derby for testing. No 59104 remained at Derby for high-speed active tests on the Midland main line on 5, 7 and 9 November, to establish whether the Class 59 bogies could operate at speeds in excess of 60 mph on a regular basis. To test the ride, a special yaw damper was fitted to No 59104 which hauled DM&EE Test Car No 6 plus five stored Mk III sleeper vehicles from Derby to Cricklewood and back on 5 and 7 November. The photograph shows the test train passing Saxby on 5 November, having been halted at Melton Mowbray for 45 minutes to rectify a minor traction motor fault. The locomotive carried neither name nor number for these tests.** ***Colin J. Marsden***

Acton: **The arrival of ARC's Class 59 locomotives led to a number of timetable changes. Two trunk trains are now booked to run from Whatley to the London area, one conveying portions for Allington and Chislehurst (or Tolworth when required) and the other conveying portions for Dagenham and Ardingly. No 59101 descends Acton bank on 4 December with 6V17, the 10.00 Allington-Whatley 'empties'. All trains between the West London line and the Western Region main line were re-routed via Willesden South West sidings and Acton bank when the direct link at Old Oak Common was removed in late October.** ***Brian Denton***

NEW SERVICES

Witton-Whitwell: **The new flow of limestone from Whitwell to Witton was incorporated into the timetable in July 1990. The train plan was devised in two parts, using Toton yard as a pivot. This enabled locomotives to work 'out and back' diagrams from their home depot, with no wasteful light engine movements. Trains were scheduled to run from Toton to Witton and back on Monday, Wednesday and Friday mornings, and from Toton to Whitwell and back on Tuesday, Thursday and Friday evenings. An overnight stop in Toton yard was scheduled for loaded trains on Tuesdays, Thursdays and Fridays (ie over the weekend) and for the return 'empties' on Mondays and Wednesdays. All trains were routed via the Sutton Park freight-only line on the outskirts of Birmingham. Construction-liveried Class 56 No 56064 passes the former Post Office sidings at Sutton Park on 30 July with 6D08, the 14.23 Witton-Toton 'empties'. The train comprises 23 POA wagons on lease from Tiger Rail.** ***Paul Shannon***

Whitwell quarry: **The loading procedure at Whitwell is unsophisticated but efficient. No 56058 has just propelled its train of empty POAs into the quarry siding on Friday 17 August and loading is already under way. A couple of hours later the train will be ready to depart for Toton where the wagons will be recessed until Monday morning.** ***Paul Shannon***

Angerstein Wharf: Class 56 No 56037 *Richard Trevithick* backs into the recently opened Bardon discharge terminal at Angerstein Wharf with 6Z55, the 23.20 service from Merehead, on 9 August. After discharging its load of limestone from the Mendips, the train will proceed to Bardon Hill, Leics, in order to collect a load of granite for Bardon's other receiving terminal at Thorney Mill. The sharing of traction and rolling-stock between different flows such as these is becoming increasingly common on today's freight railway. The Angerstein Wharf branch also carries up to three daily trainloads of sea-dredged sand and gravel on behalf of United Marine Aggregates, whose loading facilities are visible on the right of this photograph. *Paul Shannon*

Luton Limbury Road: The late 1980s saw the conversion of a number of former coal concentration depots into aggregates-receiving terminals. One of the most recent conversions was at Luton Limbury Road, where a Tarmac terminal was set up to receive granite from Stud Farm and gravel from Marks Tey. Trains started running in 1989 but were not incorporated into the timetable until January 1990. Class 56 No 56063 *Bardon Hill* draws forward during discharge at Limbury Road on Friday 3 August, having worked 6C40, the previous day's 21.47 Stud Farm-Limbury Road. The train will depart from here as 6Z31, the 16.02 Limbury Road-Marks Tey, so that the wagons are in position for the Monday morning gravel train from Marks Tey to Mile End. *Paul Shannon*

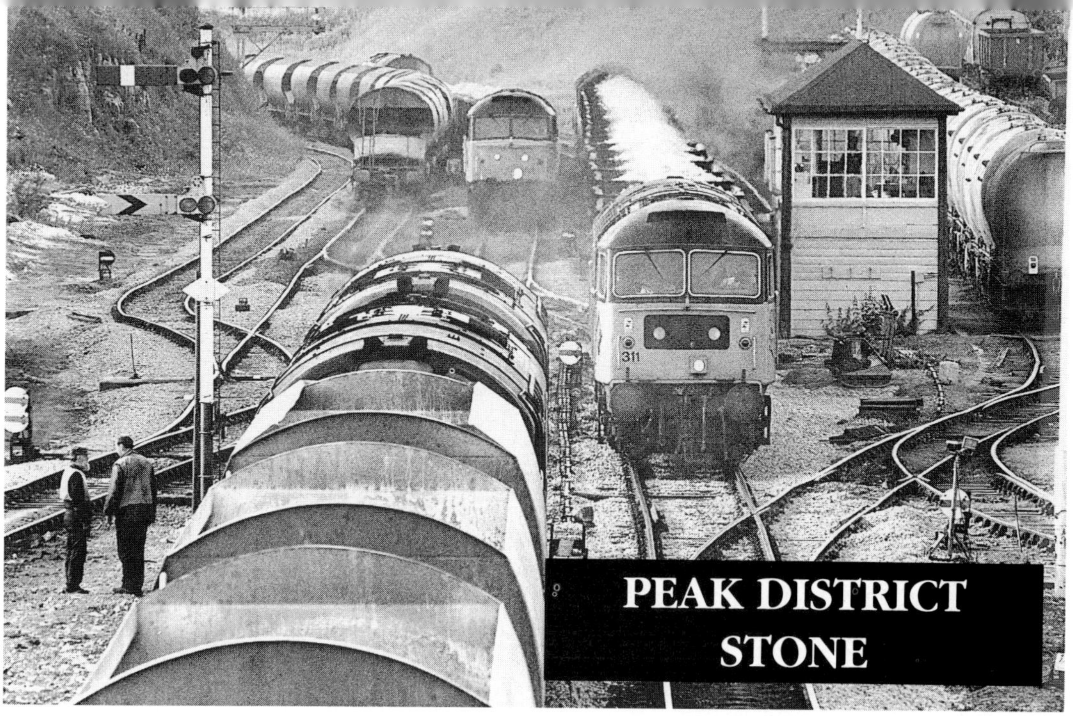

PEAK DISTRICT STONE

Peak Forest: The thriving nature of Railfreight operations in the Peak District is exemplified in this view of Peak Forest on the morning of 3 July. Class 47 No 47311 struggles away from a signal check with the 7F49 Tunstead-Oakleigh limestone train. It and its load of 50-year-old JGV (formerly PHV) hopper wagons are banked up the incline from Tunstead by Class 47 No 47599. Once past the turn-out in front of No 47311, the banker will drop off the Oakleigh train and reverse into the yard on the right in order to pick up the PCA cement wagons which are bound for Earles Sidings. To the left is Class 47 No 47290 which awaits departure with the 10.21 to Leeds. This too will receive banking assistance, provided by the Peak Forest pilot, Class 37/5 No 37688. Filling the last remaining siding accessible from the main line, visible on the far left, are the wagons which will form the morning service to Bletchley. Because of this overcrowding, Class 37/5 locomotives Nos 37686 and 37684 must wait at the signal in the foreground before they can enter the yard with the morning 'empties' from Salford Hope Street. *Michael Rhodes*

Hindlow: Since January 1988 the Hindlow branch has been scheduled to receive two daily trainloads of limestone from Tunstead, Saturdays and Sundays included. The stone is required to feed ICI's lime works at Hindlow, now that ICI's quarrying activities at Hindlow have ceased. Class 37/5 locomotives Nos 37687 and 37677, painted in 'new' and 'old' Railfreight liveries respectively, arrive at Hindlow on 27 January with 7T81 from Tunstead. The service uses ICI's own fleet of hopper wagons, which also operate from Tunstead to Lostock, Oakleigh, Bredbury and Collyhurst Street. *Paul Shannon*

Tunstead-Widnes: **Flows of limestone from Tunstead to Tarmac receiving depots at Pendleton and Widnes ceased in early 1990, although Trainload Construction was hopeful that the traffic might start up again at some point in the future. Class 47 No 47473 passes Monks Siding signal box, between Fiddlers Ferry and Warrington, with the 08.18 'empties' from Widnes (train number 6H43) on 10 August 1989. The train would continue its journey back to the Peak District via Warrington Arpley (reverse), Northwich and Northenden. The wagons are in storage at the time of writing at Tarmac's repair depot at Topley Pike, between Tunstead and Buxton.** ***Paul Shannon***

THE REDLAND SELF-DISCHARGE TRAIN

Mountsorrel-Banbury: **'Large logo'-liveried Class 56 No 56103 passes Cossington on the Midland main line with 6M31, the 11.41 Mondays-only Banbury-Mountsorrel 'empties', on 23 April 1990. The discharge wagon is clearly visible at the head of the formation, followed by three ten-vehicle sets of hopper wagons. The Mountsorrel-Banbury flow was scheduled to operate on Mondays throughout the year, with additional Thursday workings (in earlier timings) when required.** ***Paul Shannon***

TRACTION CHANGES

Colchester Hythe: In February 1990 the Redland self-discharge train made its first visit to the Clacton-on-Sea branch with a consignment of Leicestershire stone for Hythe coal depot. Class 56 No 56060 is pictured at Hythe on 24 February whilst discharge is in progress. *Michael J. Collins*

Marks Tey: The Construction sub-sector extended Class 56 traction to a number of train services in the London area during 1990. Sand trains from Marks Tey to Mile End had for many years been hauled by a pair of Class 31s, but from spring 1990 onwards a single Class 56 was rostered. On 23 May, the first day of the new arrangements, Class 56 No 56063 *Bardon Hill* prepares to leave Marks Tey sand siding with 6C30 to Mile End. Apart from the first wagon, which belongs to RH Roadstone, the train is allTiphook bogie hoppers. *Michael J. Collins*

Cricklewood-Forders Sidings: By the end of 1990 all household refuse trains in Southern England had gone over to Class 56 haulage, enabling further reductions to be made in the pool of Class 31 locomotives based at Stratford for Trainload Construction services. Class 56 No 56048 is pictured near Kempston, on the Bedford-Bletchley line, with 6A61, the 08.15 Cricklewood-Forders Sidings refuse train. *Paul Shannon*

Grain-Shakespeare Cliff: A milestone was reached on 8 November when the 1,000th train of concrete segments for Channel Tunnel construction left the TransManche Link terminal at Grain. Needless to say, the opportunity for celebration was not missed. A pair of immaculate Class 33s, Nos 33021 and 33051 *Shakespeare Cliff,* were provided to haul the special train from Grain to Shakespeare Cliff, running as 6Z85, the 12.43 departure from Grain, seen surrounded by a sea of segments at Grain. *Brian Morrison*

Cliffe-Sevington: Blue-liveried Class 33 locomotives Nos 33004 and 33058 pass Albany Park with 6C66, the 10.02 Sevington-Cliffe 'empties'. This service conveys sea-dredged aggregates from the river Thames to the Channel Tunnel construction site and uses hired Tiphook bogie hopper wagons. At one stage it was intended to replace the pairs of Class 33s on Channel Tunnel construction trains with single Class 60 locomotives, but the delays which have occurred in commissioning the Class 60s appear to have frustrated these plans. *Peter Tandy*

CEMENT TRAFFIC

Northfleet-Theale: **The only regular and long-term flow of cement carried by rail from Northfleet in 1990 was to Theale, where Blue Circle has a major distribution terminal. Class 33 locomotives Nos 33009 and 33029 pass Wokingham with 6O60, the 17.48 Theale-Northfleet 'empties' on 25 May. This daily train could be supplemented when required by an additional departure at 14.53, routed via the Western Region main line and Kensington Olympia, but the opportunity was rarely, if ever, taken up.** ***Peter Tandy***

Birmingham Curzon Street: **At the reopened Curzon Street cement terminal in the heart of Birmingham, Class 08 shunter No 08832 shunts empty PCA cement tank wagons on the morning of Monday 30 July. The wagons will be collected shortly by a Class 37/5 locomotive and taken back to Padeswood (Penyffordd) cement works as 6R63, the 12.35 Curzon Street-Penyffordd.** ***Paul Shannon***

Northenden: One short-haul cement flow where Railfreight retains an involvement is from Earles Sidings (Hope) to Northenden. Indeed, the permanent way at Northenden cement terminal was being refurbished when this photograph was taken on 23 August. Cement reaches Northenden by means of an Earles Sidings-Peak Forest trip working which feeds into 6F17, the 14.28 Peak Forest-Warrington service. This view shows 6F17 calling at Northenden to exchange loaded wagons for empty ones, the 'empties' being circulated via Warrington. Class 47 No 47445 is in charge. Although 6F17 is sponsored by Trainload Construction, it may also carry wagonload traffic on behalf of Railfreight Distribution, such as fuel oil for Buxton motive power depot and lime from Hindlow. On this occasion one loaded lime hopper and one empty oil tank can be seen waiting on the main line whilst No 47445 completes its shunting manoeuvres. *Paul Shannon*

Grangemouth: When Railfreight Distribution withdrew its operations from Millerhill yard and the east coast of Scotland in autumn 1989, the bulk sub-sectors clubbed together to provide a daily service between Millerhill and Inverness. This service continued to run successfully during 1990, with motive power provided by a Class 37/7 locomotive from the Eastfield Petroleum pool. The train calls at Grangemouth in both directions, and the principal traffic flows catered for are cement from Oxwellmains to Grangemouth, cement from Oxwellmains to Inverness, liquid petroleum gas from Grangemouth to Inverness, diesel fuel oil from Grangemouth to Inverness, and household coal from various English sources to Inverness. On 30 August the afternoon southbound working, 6B70, the 13.20 Inverness-Millerhill, approaches Grangemouth hauled by Class 37/7 No 37707 and Class 47/7 No 47709. *Paul Shannon*

Powderhall: The first household waste by rail scheme in Scotland came to fruition at the end of 1989, following the successful schemes in Greater London, Greater Manchester and Avon. Track was relaid on the former Granton branch as far as Powderhall, where containerised refuse is gathered from the whole of the Edinburgh area. At Kaimes, on the Edinburgh-Shotts-Glasgow line, a siding was provided to enable the refuse to be transferred to a disused quarry. A daily service is operated between Powderhall and Kaimes, with up to 11 container-carrying wagons hauled by one of Trainload Construction's small pool of Class 47 locomotives based at Eastfield depot. No 47004 is pictured running round its train of empty containers at Powderhall on 31 August, after working 4B44, the 10.15 Kaimes-Powderhall. *Paul Shannon*

Northolt-Calvert:* Trainloads of refuse from Northolt to Calvert are normally conveyed during the hours of darkness, but resignalling work on the Chiltern lines meant that several trains were rescheduled to run during the daytime in summer 1990. Class 56 No 56060 passes through the rolling landscape between Beaconsfield and High Wycombe with 6Z70, the 14.50 Northolt-Calvert 'special', on Monday 13 August. *Paul Shannon

Hoo Junction-Forders Sidings:* In contrast to the Class 47s used on the temporary waste trains from Barking and the pairs of Class 31s used on those from Willesden, the new trains from Hoo Junction to Forders Sidings carrying waste from Chatham were entrusted to Class 56 traction from the outset. Class 56 No 56078 arrives at Forders Sidings on Saturday 22 December with 6M66, the 06.30 departure from Hoo Junction. In general, daytime trains during the week are routed via the Midland main line with reversal at Bedford, whilst nighttime trains plus those on Saturdays are routed via Willesden and Bletchley. The Bedford route is awkward in that it involves a reversal, but fewer signal boxes need to be open on the Bedford-Bletchley line if trains approach from Bedford than if they approach from Bletchley. *Paul Shannon

Chatham-Hoo Junction:* A single Class 33 locomotive is provided to haul up to nine daily trainloads of containerised waste from Chatham Dockyard to Hoo Junction, where the traffic is remarshalled into three longer trains for Forders Sidings, Stewartby. Class 33 No 33046 passes Gillingham with 6D90, the 11.15 Chatham-Hoo Junction service on 19 October 1990. *Brian Morrison

Chapter 3

TRAINLOAD METALS

In common with the other bulk sub-sectors, Trainload Metals suffered as a result of the general downturn in the economy during 1990. Almost all the markets for finished steel recorded a decline in the latter half of the year, with previously buoyant industries such as automotive, engineering and construction all having to reduce their level of activity. Trainload Metals was to some extent shielded from the downturn because its main carryings are in the raw and semi-finished steel categories, and British Steel did not immediately cut back the manufacturing of raw steel, preferring instead to stockpile the product. Given that stockpiles cannot grow indefinitely, however, Trainload Metals was having to face the likely prospect of further reductions in traffic towards the end of 1990 and into the early part of 1991.

Whilst the fortunes of any transport concern will generally reflect the fortunes of the industries served by it, there may be instances where the contraction of an industry actually leads to a greater requirement for transportation. Back in the early 1980s, for example, Railfreight gained significant flows of semi-finished steel traffic as a direct result of the closure of blast furnaces at certain key steel works such as Corby and Shotton. During 1990 the decision was taken to close the hot strip mill at British Steel's Ravenscraig works, with a target date of March 1991. Whilst on this occasion an actual increase in traffic would be unlikely, Trainload Metals would benefit to the extent that the discontinued flows of hot coil from Ravenscraig to Shotton, Ebbw Vale and Trostre would be replaced by a new flow of steel slabs from Ravenscraig to South Wales. The changeover from coil to slab traffic would be a gradual one; indeed, during the latter half of 1990 some of the scheduled trains from Ravenscraig were already carrying slabs as well as or instead of coil. In October 1990, however, services on the principal West Coast axis were completely recast. The old 'triangular' system where certain trains ran loaded from Ravenscraig to South Wales, loaded from South Wales to Shotton and empty from Shotton to Ravenscraig gave way to a simpler one in which all trains ran to or from Shotton, generally loaded both ways between Shotton and South Wales and loaded southbound only between Shotton and Ravenscraig. The new and old systems are illustrated in the diagrams overleaf.

Also with effect from October 1990, Trainload Metals reassessed its traction requirements for West Coast steel traffic. Traditionally trains on the northern half of the route had been electrically hauled between Warrington (or Crewe) and Mossend, using a variety of AC traction drawn from a common pool. From the October timetable change, full fleet dedication was introduced for AC electrics, and the only Railfreight sub-sector to retain such traction on its books was Distribution. It would have been possible for Trainload Metals to acquire a small fleet of electrics for its West Coast services, or for the sub-sector to borrow suitable traction from Distribution, but in the event it became clear that the efficiency of the operation could be improved by not using electric traction at all. The increased power consumed by diesel as distinct from electric locomotives

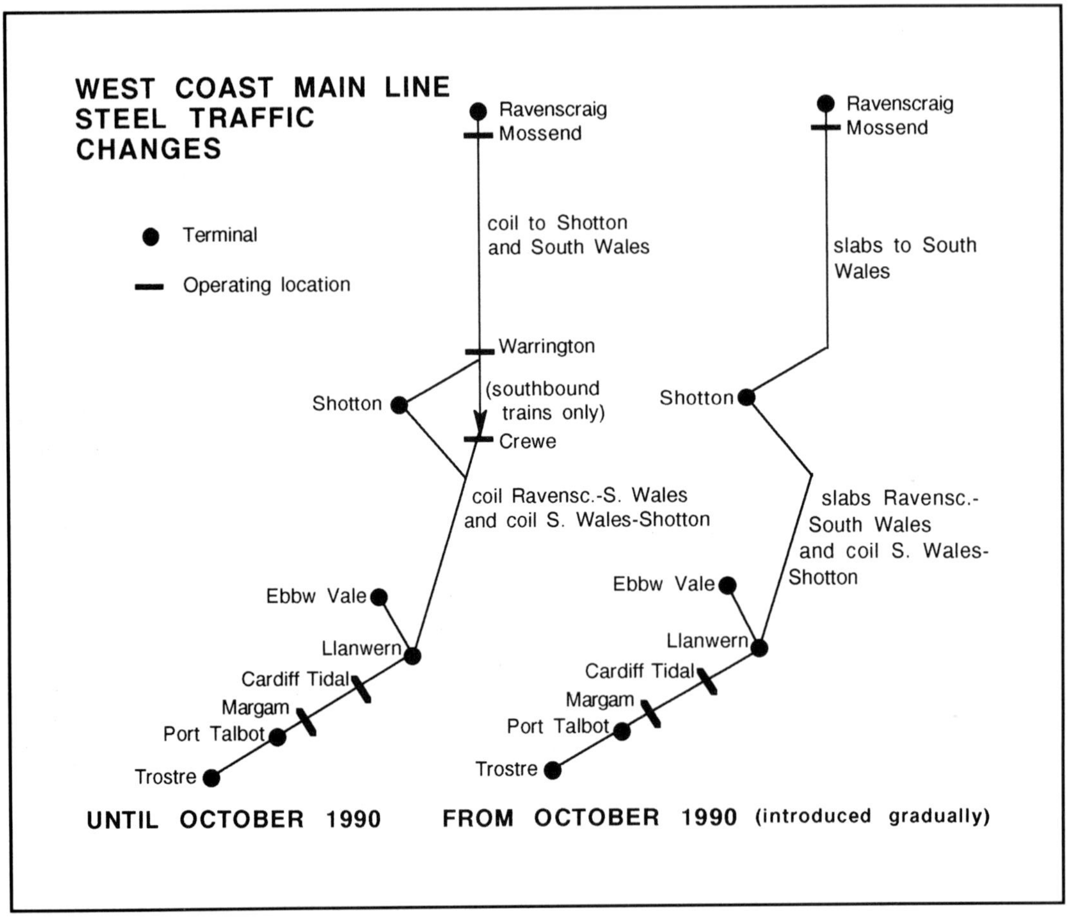

on the undulating West Coast Main Line would be amply compensated for by the savings in manpower costs and waiting time achieved by no longer changing locomotives and crews at Warrington, not to mention the uncertainty which might arise if borrowed traction were used. The type of traction adopted as standard from October, then, was the class 37. Between Shotton and Mossend or Ravenscraig, trains were now hauled by pairs of unrefurbished Class 37/0s or refurbished Class 37/5s, allocated to Motherwell depot. To cater for the new duties, Motherwell's fleet was augmented by eight Class 37/5s from Thornaby. It was hoped that Motherwell's Class 37/5 fleet would be further increased to a total of 13 in early 1991, displacing more Class 37/0s off the West Coast Main Line. Between Shotton and South Wales, trains continued to be hauled by a single refurbished and ballasted Class 37/7 or Class 37/9 locomotive from the existing Metals stud at Cardiff Canton.

Trainload Metals was able to improve the efficiency of its operations in Yorkshire by further restricting its use of marshalling yards. Scunthorpe West yard was closed completely, with most of its traffic being transferred to the nearby Anchor exchange sidings. Scunthorpe West yard will be remembered as British Rail's last operational hump yard, but the hump was rapidly becoming an anachronism in an age when most traffic travels in block trains and when more and more wagon types are barred from this type of shunting. Anchor exchange sidings are situated adjacent to the main 'C' entrance to Scunthorpe steel works, and are ideally situated as a staging point for outward consignments of finished and semi-finished steel. Over in Sheffield rumours were rife that Tinsley yard was to be closed, following the loss of Railfreight Distribution traffic there in October 1989. In fact, Trainload Metals will retain a small two-ended yard at Tinsley, mainly to cater for traffic to and from the SMACC and SPACE factories which lie nearby and have no marshalling facilities of their own. It is Tinsley's role as a gathering point for Sheffield scrap traffic and as an interchange point for general long-distance traffic which will cease. Most of the work will be transferred to Aldwarke and Tees during 1991.

Traditionally Railfreight has carried a high proportion of raw materials for the steel industry, but a much lower proportion of the industry's finished products. This is because finished steel needs to be transported in relatively small quantities to a large number of different destinations. That said, flows of finished steel to the West Midlands terminals were healthy during 1990, and an increase was recorded at one or two other locations such as the Fogartys depot at Blackburn. Several railway-owned handling terminals were closed during the year, however. At Southampton, the entire goods yard at Bevois Park was closed, but the steel traffic was retained and transferred to a more compact site on the other side of the main line. At Leeds, traffic to the BR terminal at Whitehall Road had been declining for some time, and closure was seen as inevitable. Two other privately-owned steel terminals continue to operate in Leeds, plus one at nearby Wakefield. In South Yorkshire the decision to close the relatively busy Sheffield Freight Terminal came as a surprise, but most of the traffic formerly handled here would stay on the railway and would be transferred to the expanding facilities at Rotherham Masborough.

Large quantities of imported steel and steel for export continue to travel by rail. The major success story for 1990 was Boston Docks, where dwindling traffic by the wagonload was transformed into a daily block train of imported steel for the West Midlands. Boston had traditionally been one of the remoter outposts of the Speedlink network, providing modest quantities of steel and other commodities, but the Speedlink connection was withdrawn in January 1990 as part of Railfreight Distribution's rationalisation programme. However, Trainload Metals entered into negotiations with Boston shipping agents Read & Sutcliffe, with very positive results. Not only was the imported steel traffic retained, but the tonnage was increased almost fourfold to warrant a daily block train to Round Oak and Brierley Hill steel terminals in the West Midlands. The new service allowed a 24-hour turn-round of wagons, much better than had been possible on the old Speedlink routeing via Doncaster and Bescot yards.

Elsewhere in the country, buoyant steel traffic from Scunthorpe to the continent via Dover warranted the operation of a number of block trains, though these were not considered permanent enough to be included in the timetable. Other export flows which operated during the year were from Teesside to Seaham and from Sheerness to Chatham, along with occasional imported steel movements from Chatham docks to the West Midlands. On the debit side, wagonloads of imported steel from Kings Lynn to the West Midlands were expected to cease by the end of the year, since tonnages had become too low for the service to remain viable.

The traction resources of Trainload Metals remained relatively constant, with large-scale changes not expected to occur until the successful entrance into service of the Class 60 fleet. When the situation was reviewed in late 1990, Metals was expecting to receive 16 Class 60s, comprising ten out of the first 16 plus six at a later stage in the building programme. The majority of the Metals Class 60 fleet would be based at Thornaby, though minor servicing would also be undertaken at Cardiff, Margam and Motherwell. Those Class 60s used on Immingham-Scunthorpe iron ore trains would, however, be based at Immingham together with Immingham's much larger Petroleum fleet. The cascade of motive power resulting from the arrival of the

Class 60s would enable Trainload Metals to eliminate Class 20s and unrefurbished Class 37/0s from its services. This process would be further hastened by the transfer of a number of Class 56 locomotives to the Metals sub-sector, provided that trials were successful. A long-term requirement was still foreseen for the refurbished Class 37/5, 37/7 and 37/9 locomotives which currently belong to Metals.

The main rolling-stock development of the year came to fruition on 28 February, when Trainload Metals took delivery of the first of 12 prototype steel-carrying wagons built by the private manufacturer RFS Engineering. The order comprised four vehicles each of three different designs, with lengths of 19m, 16m and 13m respectively. Each type was intended to carry a variety of loads, although Metals managers are mindful of the fact that versatility has to be weighed against effectiveness in carrying one commodity. For example, a wagon might lose some of its coil-carrying capacity if it was built to carry steel plate as well, and this would be an unnecessary restriction if the wagon concerned was expected to spend 80 per cent of its time carrying coil. Two of the new designs, those with lengths of 16m and 13m, can in fact carry either plate or coil, thanks to the provision of a deep well with removable bolsters across it. All the wagons have a welded steel underframe and can be fitted with a flexible weatherproof cover for the conveyance of bright steel if required. All are designed to international standards which means that they will be capable of operating to mainland Europe via the Channel Tunnel. A firm order for a new fleet of wagons would only be placed after exhaustive testing of the prototype designs. For up to six months after delivery, then, the prototypes were placed in service trains and visited each major steel works in turn, partly to test the robustness of each design and partly to make sure that Railfreight had the right product for the customer.

The railway-owned fleet of steel-carrying wagons was increased in April 1990 when the 181 scrap metal carriers formerly operated by Railease were purchased by Trainload Metals. The remainder of the Railease wagon fleet had already been transferred into Tiphook ownership. All bar one of the 181 wagons were built between 1982 and 1984, and their primary duty has always been to carry scrap metal to the two United Engineering Steels plants at Aldwarke and Stocksbridge. Upon transfer the wagons were recoded from POA to SSA, but no structural alterations were carried out. The other main consumers of railborne scrap traffic, Sheerness Steel and Allied Steel & Wire in Cardiff, continued to use privately-owned fleets of wagons. The year ended with a small number of vacuum-braked steel-carrying wagons still in service, such as BDW bogie bolsters, but these were expected to be eliminated early in the new year.

NEW SERVICES

Hardendale-Margam: The long-established flow of lime from Tunstead quarry to Margam for use in Port Talbot steelworks gave way to a new flow from British Steel's own quarry at Hardendale, near Shap Summit, during 1990. The lime continued to be carried in railway-owned CBA covered hopper wagons, hauled by a Cardiff-based Class 37 locomotive. Pulling away from a crew change at Hereford station on 9 July is Class 37 No 37278 – a Metals locomotive despite its Coal markings – with 6Z35, the 09.15 Carlisle-Margam service. From October the southbound train was retimed to run overnight as the 17.50 departure from Hardendale. *Paul Shannon*

BXA to BOA conversion: Some of the imported coil from Boston (see overleaf) is carried on BOA bogie coil wagons, such as No 990038 which is pictured left on 12 September 1990 at Boston. No 990038 was built at Shildon in 1979 as a BXA trestle wagon and only recently converted to coil-carrying configuration, a consequence of the reduction in sheet steel production and the increase in coil traffic seen in the late 1980s. Below left is BXA trestle wagon No 990020 in its original condition, photographed at Immingham in 1986. *Both Michael Rhodes*

Boston Docks: **Boston Docks nearly lost its rail connection in January 1990 when the one regular service there – a Speedlink trip working from Doncaster – was withdrawn as a cost-cutting measure. Thanks to careful consultation between shipping agents Read & Sutcliffe and Trainload Metals, however, a new daily block train was introduced carrying imported steel direct to the West Midlands steel terminals. Previously shunting at Boston had been carried out by a BR-owned Class 08 locomotive, outstationed from Doncaster at a distance of 50 miles. In addition, two full-time BR staff had been based in the docks – a shunter and a rolling-stock technician. Economies were achieved by arranging for the Port of Boston to purchase its own Class 08 locomotive, to be maintained on site, and by reducing the complement of BR staff to one. The new Boston-West Midlands service uses Trainload Metals motive power throughout. On 12 September, Class 37 No 37054 is pictured at Boston shortly after arriving light engine from Bescot. Unfortunately on this occasion it was decided that the five loaded wagons visible in the background did not constitute a worthwhile load, so No 37054 departed light again after just a few minutes. In spite of minor hiccoughs such as this, the service has been a great success and has preserved a Railfreight presence in a location which might so easily have been wiped off the map.** ***Michael Rhodes***

Brierley Hill: **Class 37 No 37255 arrives at Brierley Hill on 30 July with 6G27, the 05.51 service from Toton. This is the daily train from Boston which has been stabled at Toton overnight. The coil is being carried in a mixture of Powell Duffryn JSA steel carriers and BOA open wagons, the latter requiring the use of plastic sheeting to protect the delicate load of cold-reduced coil from the elements.** ***Paul Shannon***

SHORT-TERM FLOWS

Port Talbot-Ravenscraig: There are well-established flows of imported iron ore from Port Talbot to Llanwern, from Immingham to Scunthorpe, and from Hunterston to Ravenscraig. Occasionally, however, ore needs to be transported between sites not normally associated with each other, perhaps because of discharge problems at one of the ports, or perhaps in order to even out stockpiles. In summer 1990 there were regular trainloads of iron ore from Port Talbot to Ravenscraig, which at over 400 miles per single journey provided Railfreight with some useful business. Trains were hauled by a pair of Metals Class 37s throughout, foreshadowing the introduction of diesel haulage on all West Coast Main Line Metals services in October. Motherwell-based Class 37s Nos 37145 and 37099 pass Winwick with ten loaded JTA iron ore tipplers on 18 July. *Paul Shannon*

Ravenscraig-Lackenby: During April and May of 1990 Railfreight fulfilled a short-term contract to move 10,000 tonnes of steel billets from Ravenscraig to Lackenby for milling and rolling. This necessitated ten special freight services, one of which is seen winding its way into Tyne yard on 1 May behind Class 31s Nos 31252 and 31305. Both locomotives had been borrowed from the departmental sector to cater for the extra traffic. *Michael Rhodes*

Scunthorpe-Teesport: On 22 May 1990 Class 37 No 37110 roars past Picton with 6X09, a special from Scunthorpe Anchor yard to Tees yard. This flow of beams from Scunthorpe was bound for Teesport and the export market and is another example of Metals sub-sector traffic which may merit an extra short-term trainload service from time to time. *Michael Rhodes*

Ravenscraig-South Wales: Railfreight-liveried Class 90 No 90041 passes Winwick Junction with 6Z23, the 12.05 Mossend-Crewe 'special', on 18 July. The traffic comprises both coil and slab steel from Ravenscraig to South Wales. The first vehicle is one of the small batch of pre-production steel-carrying wagons delivered to Trainload Metals during 1991. Note how the coil is carried in a well in the wagon floor, instead of in a raised position as on the two BAA wagons behind it. From October coil traffic from Ravenscraig to South Wales was diverted to run via Dee Marsh Junction instead of Crewe, using Class 37 locomotives throughout. *Paul Shannon*

Scunthorpe West Yard (1): 1990 saw the end of hump shunting on British Rail. The last operational hump yard was at Scunthorpe. This yard was also the last to be built, dating back only to 1971. Its 19 sorting sidings were installed to handle the heavy steel traffic from the nearby Scunthorpe steelworks and ranged in length from 27 to 73 standard length units (SLUs). Here we see hump shunting in operation in March 1990. *Michael Rhodes*

Scunthorpe West Yard (2): The yard was equipped with Dowty retarders, which are visible in this photograph as little black spikes sticking up inside the running rails. These were designed to slow wagons down from a humping speed of 8.5 feet/second to 4.5 feet/second, allowing wagons to buffer into each other without damaging their cargoes. In this view, taken in 1986, a rake of SPA steel wagons winds gently down into the yard. A 10 mph speed limit was imposed throughout the yard to prevent damage to the retarders and this ensured a slow departure for all westbound freights. The yard finally closed in May 1990, just over 61 years after the first automated hump yard was opened at Whitemoor, Cambridgeshire. *Michael Rhodes*

Grimsby-Tinsley: **Stainless steel produced in Sheffield is exported to Germany in slab form where it is rolled into stainless steel coil. The coils are then returned to the United Kingdom through Grimsby, Goole and Boston. In 1990 most of the imports were handled at Grimsby, which led to a rail revival for the port and the associated yard at West Marsh. On 20 February Class 20s Nos 20046 and 20112 pass Scunthorpe Anchor yard with 6T60 from West Marsh to Tinsley. The increased coil traffic has also led to the conversion of BXA plate wagons into BOA coil carriers, several of which can be seen in this train.** ***Michael Rhodes***

Flixborough branch: **A little-known freight-only byway is the Flixborough branch which winds north from Scunthorpe to the port of Flixborough on the river Trent. On 21 March 1990 Class 08 No 08543 passes Crosby Mines with the first trip of the morning from North Lincoln Junction (Scunthorpe) to Dragonby yard. The sidings at Dragonby once acted as the reception point for traffic to Normanby Park steel works, but their only use since the closure of Normanby Park has been to 'stage' traffic to and from Flixborough. Trains on the branch run as required, depending on the arrival of boats at the quayside at Flixborough. More regular traffic may return to the erstwhile North Lindsey Light Railway in 1991 if plans to dump waste gypsum at the former ironstone quarries north of Scunthorpe come to fruition.** ***Michael Rhodes***

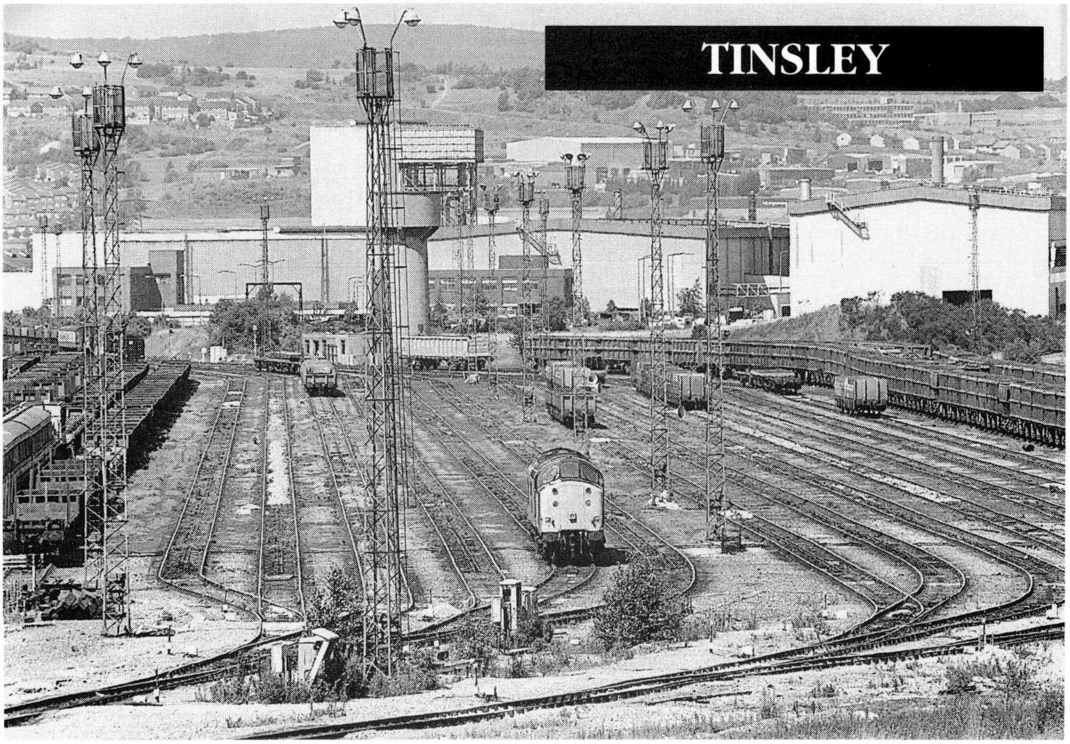

Tinsley: Opened in October 1965, Tinsley was the first large yard to be built with Dowty wagon retarders. It survived as a hump yard until December 1984. Under sectorisation, Tinsley became the primary responsibility of the Metals sub-sector, serving local steel plants and also acting as an interchange point for long-distance traffic. By the end of 1990, however, plans had been drawn up to downgrade Tinsley to a terminal yard handling only traffic for the adjacent Tinsley Park (SMACC) and Shepcote Lane (SPACE) plants. Scrap metal trains which formerly headed for Tinsley would be diverted to Aldwarke, adjacent to the UES works which is a major consumer of railborne scrap, and Aldwarke would become the base for trips to UES's other plant at Stocksbridge. In this general view of Tinsley, on 1 August 1990, Class 37/7 No 37717 awaits the arrival of 6V64, the 09.45 Tees-Margam service. The main yard is almost empty, with the sidings on the left being used to store condemned rolling-stock and those on the right holding many of Railfreight's newly acquired SSA scrap wagons. *Michael Rhodes*

Ravenscraig (1): The decision to cease producing rolled coil here will not initially affect the influx of raw materials. Plans to continue steel production until at least 1993 mean that flows of iron ore, coal and limestone are safe for at least two years. On 20 June 1990 Class 37s Nos 37049 and 37051 arrive at Ravenscraig No 4 yard with 6D11 from Hunterston ore terminal. The first Class 60s appeared on these trains in December 1990 (see page 123). *Michael Rhodes*

Ravenscraig (2): Sentinel 0-6-0 diesel-hydraulic engine No 41 pulls out from underneath Ravenscraig's blast furnaces with a train of molten slag in old-fashioned cauldron wagons. This slag will be dumped a mile north of the steel works on open ground. *Michael Rhodes*

Ravenscraig (3): Class 26 No 26042 provides unusual motive power for the 6T09 trip from Mossend yard to Ravenscraig No 3 and No 1 yards. The train is seen here reversing into No 1 yard with the daily delivery of lime from Shap. The diagram is usually handled by a Class 08 locomotive but on 20 June the class 26 was seconded from the 6T14 Clydesdale trip because of reduced activity at Clydesdale. On 9 November it was announced that BS Clydesdale was to close with the loss of 1,200 jobs. The plant had apparently been losing money for six years as the demand for its pipes fell due to reduced activity in the North Sea oil and gas fields. *Michael Rhodes*

THE METALS NETWORK

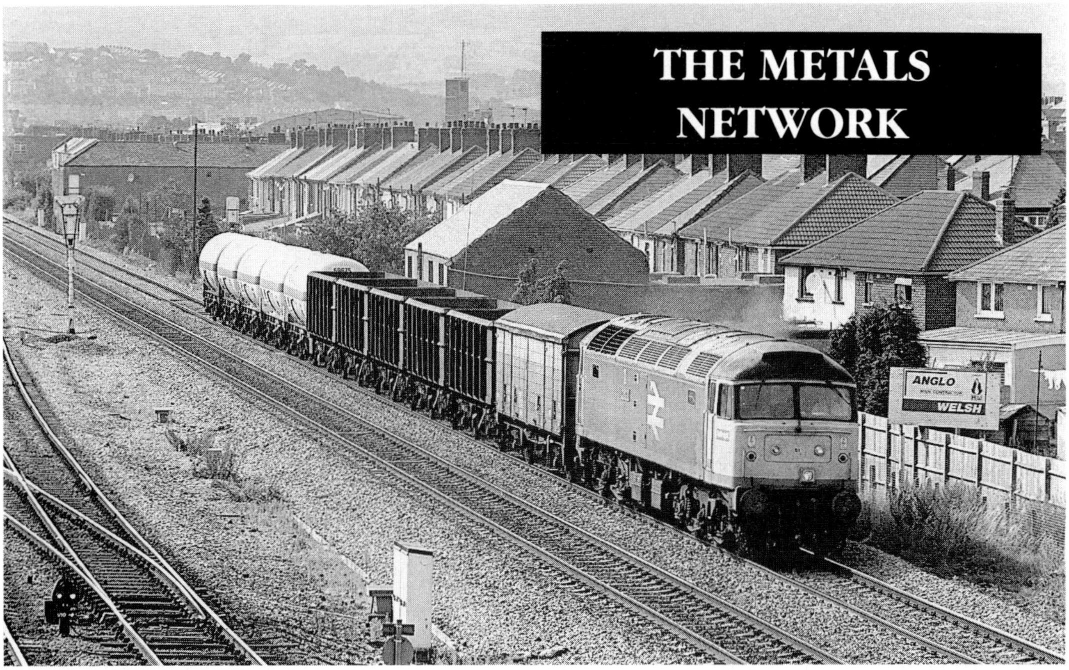

Cardiff-Willesden: **Inter-sub-sector cooperation is the name of the game on this freight service seen passing East Usk on 8 August 1990. Class 47 No 47142 is heading 6M50, the 15.20 Cardiff Tidal-Willesden Brent Metals service. The Metals sub-sector traffic is the five empty POA scrap wagons in the centre of the formation, and the non-Metals traffic comprises one ZRA departmental van and five TTA gas tanks. The mutual benefits of shared freight services have been well exploited in Scotland and will no doubt spread elsewhere as the role of Speedlink declines. *Michael Rhodes***

Sheerness-Washwood Heath: **Class 47 No 47256 heads north near Leighton Buzzard with 6M42, the 08.49 Sheerness-Washwood Heath Metals train, on 3 August 1990. This train was diverted away from the Western Region to run via the West Coast Main Line in March 1990. Like a number of Metals services conveying general traffics, 6M42 is diagrammed for haulage by a Railfreight Distribution Class 47 locomotive. The consist on this occasion is typically varied and includes SPA and BDA wagons carrying slabs from Sheerness together with empty JXAs returning to the West Midlands. *Paul Shannon***

Stranraer-Tees: The wheel of sub-sector dedication has turned full circle here. Finished steel for export to Northern Ireland was first conveyed to Stranraer by a block train from Teesside in the early 1980s. The traffic was later absorbed into the Speedlink network, but regained the status of a block train in 1989 as Trainload Metals began to reduce its involvement with Speedlink. Class 37/5 No 37501 *Teesside Steelmaster* draws to a halt in the up loop at New Cumnock with 6E10, the 08.48 Stranraer-Tees 'empties', on 1 September 1990. *Paul Shannon*

Southampton Down Yard: Trainload Metals established its own terminal to handle incoming steel traffic at Southampton following the closure of the general-purpose BR freight depot at Bevois Park. The new terminal is situated on the down side of the main line, just south of Bevois Park. Loaded wagons from South Wales are detached from 6O45, the 02.50 Cardiff Tidal-Hamworthy service, and the 'empties' are returned on 6V99, the 14.53 Hamworthy-Cardiff Tidal. 6O45 is pictured setting off from Southampton Down Yard on 14 August, headed by Class 37/9 No 37901 *Mirrlees Pioneer.* Five loaded BDAs have just been detached here at Southampton, leaving ten empty BDAs and two empty BNXs in the consist for loading at Hamworthy Docks. *Paul Shannon*

Hamworthy: **The port of Hamworthy provides Trainload Metals with a steady flow of imported steel traffic. Various types of finished steel are conveyed from Hamworthy on the four-times-weekly block train to Cardiff, which connects into another Metals service from Cardiff to the West Midlands steel terminals. A Class 09 shunter is provided for hauling wagons along the quayside at Hamworthy. On 14 August, No 09025 has just hitched up to a loaded BDA wagon which it will haul as far as Hamworthy exchange sidings.** ***Paul Shannon***

Warrington Dallam: **Situated rather appropriately on the site of a former steelworks, Warrington Dallam yard is the principal railway-owned steel terminal for North West England. Traffic is conveyed as far as Warrington Arpley or Walton Old Junction yards on scheduled Metals services and is then tripped to Dallam by one of Warrington's resident Class 08 pilot locomotives. A visit on the afternoon of 10 April found Class 08 No 08809 busily sorting loaded and empty wagons at Dallam, before returning to Arpley and Walton Old Junction with the 'empties'. Still awaiting unloading were some BDAs with girders from Scunthorpe and some SKAs (converted from SPAs) with reinforcing coils from Cardiff.** ***Paul Shannon***

Cardiff-Tees: The combination of Freightliner vehicles and traditional air-braked wagons has been much publicised because of the new Railfreight Distribution services. The Metals sub-sector has long operated such trains in order to facilitate the delivery of aluminium to Rogerstone in South Wales. On 4 August Class 37 No 37718 speeds past Rumney near Cardiff with 6E88, the 20.05 Cardiff-Tees service. The first three wagons are conventional 60 mph stock whilst the rest of the train is made up of a Freightliner set returning to Lynemouth in Northumberland for another load of aluminium. *Michael Rhodes*

SCRAP METAL

Ridham Dock-Sheerness: One of the shortest-distance scrap metal flows operated by Trainload Metals is the flow from Ridham Dock to Sheerness on the Isle of Sheppey. These trains were handed over to Class 09 haulage in 1990. A well-laden trip working from Ridham Dock is pictured as it crosses the Kingsferry Bridge at Swale on 16 July, hauled by Class 09 No 09021. *Brian Morrison*

Shotton-Llanwern: **Unusual freight services may go unnoticed by the casual observer. Here on 30 March 1990 Class 37 No 37264 passes East Usk with a rake of MDV wagons , but was in fact the motive power for 6V42, the 20.40 Dee Marsh Junction-Llanwern-Swindon service. However, in the absence of empty coil wagons from Shotton, the locomotive was used to convey eight wagonloads of scrap from Shotton to Llanwern instead. The train code has been amended to 7Z42 – '7' because the MDVs can only run at 45 mph, and 'Z' to denote a special working. Such one-off trains are an excellent way of utilising spare resources.** ***Michael Rhodes***

Shipley: **Rail-served scrapyards declined greatly in number during the 1980s. Where tonnages were small or uneven, or where distances were short, Railfreight was unable to compete effectively with road haulage. Those scrapyards which retained their rail connections were mostly large-scale installations with specialist facilities, dispatching their traffic to one of the rail-fed steel plants at Aldwarke (United Engineering Steels), Stocksbridge (also UES), Sheerness (Sheerness Steel), Cardiff (Allied Steel & Wire) or Clydesdale (British Steel). At the Crossley Evans scrapyard at Shipley on 16 July, Hunslet 0-4-0 shunter** ***Prince of Wales*** **propels a single SSA wagon into the departure siding, ready to be coupled up to Class 37 No 37054. The Class 37 will wait for a second loaded SSA wagon to be shunted into this siding and will then depart as 6J45, the 14.16 Shipley-Tinsley. From Tinsley the SSAs will complete their journey to Aldwarke by means of a local trip working, a complication which will be eliminated in 1991 when Aldwarke becomes the main yard for Sheffield area scrap traffic.** ***Paul Shannon***

NORTH EASTERN MISCELLANY

Seaham Harbour: **The rail connection to Seaham harbour might have been lost in the early 1980s when rationalisation led to the loss of coal traffic down a rope-worked incline from Hetton. The BR line was retained, however, and Trainload Metals now makes regular use of it to deliver steel for export. Trains reach the harbour by leaving the main line at Dawdon and reversing at Seabanks, before descending past the shale-unloading point and down to the newly constructed warehouse. On 14 June privately-owned pilot locomotive No 4 makes just such a trip with 20 BQV wagons from Hartlepool. These were part of the first shipment of pipes to Seaham Harbour – previously beams and blooms had been handled.** ***Michael Rhodes***

Blyth-Fort William: **Class 37/5 No 37521 heads a rake of Alcan PCA wagons over the new layout at the east end of Newcastle station on 18 May. The wagons are carrying alumina from Blyth to Fort William and are being tripped to Tyne yard. At Tyne yard they will join 6S54, the 09.05 Thrislington-Ravenscraig Metals service, routed via the Tyne valley and the West Coast Main Line. Before the May timetable change, the alumina was conveyed to Scotland in a separate train, running via Berwick and Millerhill.** ***Ian S. Carr***

Chapter 4

TRAINLOAD PETROLEUM

By the end of 1990, Railfreight's petroleum carryings had stabilised at about 10 million tonnes per annum, representing an annual turnover of some £43 million. One unexpected factor which Trainload Petroleum had to contend with during the latter part of the year was the Gulf crisis. Initially this resulted in a slight increase in traffic as certain consumers added to their stocks, but in the longer term any rise in the cost of oil would inevitably lead to a decrease in rail carryings, especially where customers were able to turn to cheaper alternative sources of fuel.

The year was marked by two extensions to the Trainload Petroleum network. The first of these was the reopening of the former Staines West branch as far as Colnbrook to serve a new aviation fuel terminal for Heathrow airport. BR had had the foresight to retain the track to Colnbrook after closure to regular traffic in the early 1980s, and reopening was therefore relatively straightforward. Heathrow airport had previously received its fuel either by direct pipelines from refineries or else by a short pipeline from the Total rail terminal at Langley, and these methods of supply would continue. The main purpose of the additional terminal at Colnbrook was to give greater flexibility of supply and, in particular, to allow the smaller oil companies an independent means of access to the airport. The first train to Colnbrook ran on 1 March, bringing fuel from the Unitank storage terminal at West Thurrock on North Thameside. A weekly service was then inaugurated, using a Petroleum sub-sector Class 47 locomotive and a rake of freshly refurbished 100-tonne tank wagons. In the summer the flow from West Thurrock ceased, but a longer-distance flow from Waterston refinery was established, running up to four times weekly. The Waterston trains were hauled by a pair of Class 37s and ran in a similar path to the already existing trains from Waterston to Langley. From October it was planned to use the superior haulage capability of a Class 60 locomotive to operate one long train each day from Waterston to the London area, carrying traffic for both Langley and Colnbrook. In practice, however, the delay in commissioning the Class 60 fleet meant that the two trains continued to run separately. At the end of the year there was still the possibility of trains running to Colnbrook from North Thameside again. If this came to fruition the North Thameside trains would be timed to reach Colnbrook in the early afternoon, after the departure of the Waterston empties.

The second extension to the network was a new flow of liquid petroleum gas (LPG) from Furzebrook to Avonmouth. From 1979 until 1990 the BP terminal at Furzebrook dispatched regular trainloads of crude oil from the Wytch Farm oilfield to Fawley refinery. This flow ceased in the summer when the oil company commissioned a direct pipeline to Hamble to cater for a planned sixfold increase in output. One effect of this increase, however, was that the quantity of LPG produced alongside the oil would rise as well. Previously the small amount of LPG produced from Wytch Farm was absorbed by local customers, but now BP needed to find markets further afield. For that reason it was decided to set up a specialised distribution terminal at Avonmouth, from where the LPG would be forwarded by road to customers in Wales and South West England and by sea to ports in Britain and Europe. Railfreight was chosen as the most efficient means of transporting the LPG as far as Avonmouth. The existing rail facilities at Furzebrook were converted during the autumn to handle LPG instead of crude oil, whilst at Avonmouth BP invested £17 million in an entirely new terminal. The first train from Furzebrook to Avonmouth ran on 27 November, and the service frequency was expected to build up quickly to two trains each weekday plus one on Saturday. Class 47 motive power was used initially, but Class 60s were to be drafted in as soon as these became available. One major obstacle on the most direct route from Furzebrook to Avonmouth – via Westbury and Bristol – is Filton bank. A Class 60 locomotive can handle Filton bank easily, but not a Class 47, and it was therefore decided to send all loaded trains over the longer but more gently graded route via Basingstoke and Swindon for as long as Class 47s remained in charge.

During 1990, Trainload Petroleum brought in a number of measures designed to make its existing operations more cost-effective. On some of the busier routes out of South Humberside it made sense to run fewer but longer trains, initially double-headed but ultimately to be hauled by a single Class 60 locomotive. The maximum tonnage of trains from Immingham to Colwick (Nottingham) and Leeds was increased to 2,200 during the autumn, and the maximum tonnage from Immingham to Langley was expected to be almost doubled from its existing level of 1,500 to 2,700 early in 1991. These three flows have an excellent future, thanks to a long contract which Trainload Petroleum signed with Total in 1988. Already in the first two years of the contract the annual tonnage from Immingham to Langley has increased from 200,000 to 650,000. On the Western Region, mention has already been made of the planned increase in train length from Waterston to the London area, and certain other flows – such as Robeston to Theale – will follow suit during 1991.

Trainload Petroleum continued to reduce its involvement with the Speedlink network during 1990. Wherever practicable, Petroleum introduced its own block trains to terminals which had traditionally been served by Speedlink. In January, for example, following the total withdrawal of Speedlink from the Westbury area, a new daily Petroleum service was introduced from Ripple Lane to Frome via Westbury, carrying bitumen from Thames Haven to the Mobil terminal at Frome. Elsewhere on the network, daily wagonload consignments of perhaps two or three wagons could be combined into a block train running once or twice a week. Flows from Thames Haven to Canterbury and from Stanlow to Swindon were converted to weekly block train operation in January and October respectively.

One area of major concern was diesel fuel for BR's own diesel depots. Quantities tend to be small, and depots tend to be scattered far and wide, making trainload operation difficult. At the end of 1990 most flows to diesel depots were still in the hands of Speedlink, but in a few cases Trainload Petroleum had found a way of servicing a group of depots with its own resources. In the London area, a daily Petroleum train was operating from Ripple Lane to Temple Mills (for Stratford) and Old Oak Common, returning via Ferme Park on two days a week and with a connecting service from Old Oak Common to Marylebone on Fridays. In Yorkshire, depots at Doncaster, Knottingley, Holbeck and Neville Hill were served by a Petroleum-sponsored train from Immingham, while in the Midlands a Saturdays-only block train was running from Stanlow to Derby Etches Park.

The last few years have seen a number of smaller rail-fed oil terminals close down. Casualties in 1990 included the LPG terminal at Rowley Regis and other terminals at Newbury and Harrogate. On a happier note, the oil company Murco invested in new railway facilities in February when it began building a new general-purpose petroleum terminal at Westerleigh, north of Bristol. To mark the half-way stage in building the terminal, a ceremony was held there on 12 June during which

Petroleum sub-sector Class 37/0 No 37220 was named *Westerleigh*. The terminal was scheduled for completion early in 1991 and will be able to handle up to 34 100-tonne tank wagons at a time.

Trainload Petroleum made further moves during 1990 towards concentrating its locomotive maintenance on Immingham depot. The pool of Class 37s and Class 47s used on North Thameside petroleum traffic was transferred from Stratford to Immingham in April, and it seemed likely that Class 47s used on Stanlow trains would move from Crewe to Immingham early in the new year. Day-to-day maintenance for the North Thameside fleet continued to be carried out at Ripple Lane, but any locomotives needing heavier maintenance or repairs would now go to Immingham. The physical transfer of locomotives to and from Immingham would take place wherever possible using scheduled Petroleum trains passing through Yorkshire, such as Ripple Lane-Weaste and Ripple Lane-Kilnhurst. For the future, Trainload Petroleum intends to use Class 60 locomotives on its heaviest and longest-distance flows. The plan at the end of 1990 was for Petroleum to receive a total of 18 Class 60s. Eleven of these would be based at Immingham for use on trains from Lindsey and Humber refineries, whilst the other seven would be based in South Wales, including two for the new Furzebrook-Avonmouth flow mentioned above. Full use will be made of the Class 60's superior haulage capability, with trainloads of 20 or even 30 bogie tank wagons becoming standard on some routes. Not all services will warrant Class 60 haulage, and Trainload Petroleum expects to retain a stud of Class 47 and refurbished Class 37 locomotives for use on less intensive service groups such as those operating out of North Thameside and Grangemouth.

NEW SERVICES

Ripple Lane-Colnbrook:* The reopened Colnbrook branch was host to a weekly train of aviation fuel from North Thameside in spring 1990. The traffic was loaded by Unitank at West Thurrock and staged *en route* at Ripple Lane yard. The standard train length was 15 100-tonne bogie tanks, hauled by a Class 47 locomotive. On 19 April Class 47 No 47373 approaches Colnbrook with 6Z83, the 05.00 departure from Ripple Lane. *Paul Shannon

Waterston-Colnbrook: **The second flow of aviation fuel to Colnbrook was both more frequent and covering a longer distance than the first. Up to four trains a week started running during the summer from Waterston refinery near Milford Haven. Traction was provided from the Petroleum Class 37 pool based at Cardiff Canton, although it was hoped to draft in Class 60s as soon as these became available. The illustration shows Class 37/0 Nos 37220 *Westerleigh* and 37294 drawing out of the run-round loop at Colnbrook on 10 August with 6Z29, the 13.05 return 'empties' to Waterston. *Paul Shannon***

Furzebrook-Avonmouth: **The new flow of liquid petroleum gas from Furzebrook to Hallen Marsh Junction (Avonmouth) commenced operation in November 1990. Up to two trains daily were scheduled to run, using Class 47 locomotives as a stop-gap measure until Class 60 traction became available. Class 47 No 47094 passes Lymington Junction in the New Forest with 6V13, the 12.24 Furzebrook-Hallen Marsh Junction service. *Brian Denton***

Immingham-Leeds: In the autumn, Class 60-length oil trains started running from Lindsey oil refinery, Immingham, to two major distribution terminals at Leeds and Nottingham. Double-headed Class 47 haulage was provided in place of the booked Class 60, and a specific 'Class 60 contingency plan' pool of Class 47s was even set up at Immingham depot. On 26 September Class 47s Nos 47380 *Immingham* and 47222 *Appleby-Frodingham* approach Barnetby with 6D71, the 10.54 Lindsey (Immingham)-Leeds oil train. The load comprises 21 of Tiphook's new fleet of 90-tonne bogie tank wagons, built in France by the firm Arbel Fauvet and now on hire to Total Oil. *Brian Denton*

Fawley-Plymouth: Railfreight Distribution withdrew all Speedlink services from the Westbury area in January 1990, and the wagonload petroleum traffic carried on these services was reorganised into block trains sponsored by Trainload Petroleum. One train which started running as a 'special' in 1989 but was included in the timetable in January 1990 was a direct service between Fawley oil refinery and Tavistock Junction (Plymouth), calling *en route* at Westbury, Bristol and Exeter. The train carries mainly fuel oil for BR depots but may also convey bitumen from Fawley to Plymouth Friary. Class 47 No 47327 is in charge of the service on 1 November, running as 6V62, the 10.44 Fawley-Tavistock Junction. The location is West Grimstead, near Salisbury. *Brian Denton*

BR FUEL OIL

Canterbury tanks: For many years block trains of heating oil have run from North Thameside to a variety of destinations on the Southern Region, but the heating oil depot at Canterbury West continued to receive its supplies by the wagonload on Speedlink services until January 1990. After the January timetable change, the Canterbury traffic was transferred to a Fridays-only block train from Redhill, which at that time was the staging point for most Southern Region heating oil trains. Sometimes the Canterbury traffic was combined with traffic for High Brooms, with portions being split and rejoined at Tonbridge West Yard. Haulage took the form of either one or two Class 73 electro-diesels, depending on the load. On 20 April, Class 73s Nos 73118 The *Romney Hythe and Dymchurch Railway* and 73105 enter Tonbridge West Yard with 6R77, the 13.18 Canterbury West-Tonbridge 'empties', ready to be combined with wagons from High Brooms and continue as 7Y57, the 15.50 High Brooms-Redhill. From October 1990 the practice of staging heating oil trains at Redhill was discontinued, and Canterbury and High Brooms traffic was worked direct from Ripple Lane with Trainload Petroleum traction throughout. *Paul Shannon*

Leeds Hunslet: In several parts of the country where BR motive power depots are closely grouped, Trainload Petroleum has been able to introduce its own services to reduce its involvement with the Speedlink network. In Yorkshire a daily train runs from Lindsey oil refinery to Doncaster and Leeds Hunslet (for Holbeck and Neville Hill), also serving Knottingley on two days a week. Petroleum-liveried Class 31 No 31207 prepares to depart from Hunslet yard on 16 July with 6D97, the 10.28 Hunslet-Lindsey service. *Paul Shannon*

Stanlow-Wigan: **A light duty for one of Trainload Petroleum's Stanlow Class 47 pool, as No 47368 *Neritidae* passes Winwick with an evening 6T68 trip working from Stanlow to Wigan Springs Branch on 30 May. Fuel for Springs Branch had previously been conveyed by Speedlink trip workings from Stanlow to Warrington and from Warrington to Wigan. *Paul Shannon***

OCCASIONAL WORKINGS

Taylors Lane:* The splitting up of the Central Electricity Generating Board into two rival companies has led to some changes in fuel supply patterns to individual power stations. Most of these changes have affected only the Trainload Coal sub-sector, but Trainload Petroleum has gained a small amount of new business in supplying fuel to previously little-used oil-fired power stations. The first train of the winter season is pictured setting back into the sidings at Taylors Lane power station, Neasden, on 30 October. The train is running as 6Z83, the 09.15 Ripple Lane-Taylors Lane 'special', hauled by North Thames Petroleum Class 37/7 No 37706. The route taken on this occasion was via Tottenham, Cricklewood and Acton Canal Wharf, with reversal at Acton Canal Wharf in order to approach Taylors Lane from the south. Neasden Junction signal box had been officially closed and all signalling removed by the date of this photograph, so special arrangements had to be made to enable the train to enter the sidings. *Paul Shannon

Perth: **Class 47 No 47233 *Strombidae* accelerates south from Perth with the 6N64 empty tank train to Grangemouth on 17 August. This rarely photographed train was scheduled to run once weekly in summer 1990. No 47233 was previously a member of the Stanlow Petroleum fleet but was reallocated to Eastfield in July 1990 for working block petroleum trains out of Grangemouth. *Michael Rhodes***

Immingham-Burn Naze: **Many industrial customers consume irregular quantities of fuel oil, depending partly on their own level of activity and partly on the relative price of oil compared with other types of fuel. During 1990 oil trains ran to Burn Naze from both Stanlow and Lindsey (Immingham) refineries, and for a time during the summer a twice-weekly service from Lindsey was programmed into the timetable. Immingham-based Class 47 No 47379** ***Total Energy*** **climbs towards Copy Pit on 17 July with 6Z70, the 12.54 Burn Naze-Lindsey 'empties'.** ***Paul Shannon***

St Helens: **Stanlow Petroleum locomotive No 47324** ***Glossidae*** **approaches St Helens Central station on 18 July with 6R53, the 06.14 Stanlow-Cowley Hill fuel oil train. The 'R' in the reporting number denotes a special working. In order to enter the siding at Cowley Hill, the train will continue to Ravenhead Junction, run round, and pass through St Helens Central station a second time. Both Ravenhead Junction and Cowley Hill retain oil sidings.** ***Paul Shannon***

Stanlow-Boulby: **An addition to the timetable in October 1990 was a once-weekly oil train from Stanlow to Boulby. The service ran occasionally during the summer as a 'special'. Class 37 No 37116 passes Picton signal box with the 6Z85 'empties' from Boulby to Stanlow on 22 May.** ***Michael Rhodes***

Scarborough: An interesting survival in 1990 was the oil siding at Scarborough, receiving approximately one train a week from Stanlow refinery for general distribution in the Scarborough area. The empty tanks from Scarborough oil siding are seen setting back into Scarborough station on 16 July, ready to return to the other side of the country as 7Z59, the 16.10 Scarborough-Stanlow. The motive power is main-line liveried Class 37/4 No 37430 *Cwmbrân,* which was one of two Class 37/4s allocated to the Stanlow pool at that time. *Paul Shannon*

A PETROLEUM MISCELLANY

Holybourne: Class 47 No 47326 stands at Holybourne crude oil terminal on 17 March, shortly before departing with 6Y51, the 13.51 Holybourne-Eastleigh service. Tonnages of crude oil from Holybourne were increasing during 1990 and extra trains were necessary over and above the three booked services per week. The timings were altered from late September so that loading took place at Holybourne in the late evening. This enabled the Eastleigh-Holybourne leg to feed directly out of the morning trip from Fawley, giving better wagon utilisation. *Brian Denton*

Weaste: Thornaby-based Class 47 No 47305 runs round its train of empty oil tanks at Weaste oil terminal on 6 July. It will depart as 6E31, the 10.26 Weaste-Port Clarence. The sidings at Weaste are a remnant of the once extensive Manchester Ship Canal (MSC) railway system, and one of the MSC shunting locomotives still in use at this location can be seen on the right. *Paul Shannon*

Sunderland: **The most northerly oil terminal served from Immingham lies in Sunderland docks. On 31 August Class 31s Nos 31233 and 31319 accelerate out of the terminal with the first half of 6D40, the 13.11 service to Lindsey refinery. Such long-distance services are likely to be candidates for Class 60 haulage during 1991.** ***Michael Rhodes***

Ashton-in-Makerfield: **One of the most lightly loaded block trains operated by the Petroleum sub-sector is the bitumen working from Immingham (Lindsey refinery) to the Kelbit private siding at Ashton-in-Makerfield. The standard train length is 11 45-tonne tanks, hauled by a single Class 31 locomotive from the Immingham Petroleum fleet. The Kelbit siding lies just off the West Coast Main Line near Golborne, and access is gained by using the remaining stub of the Haydock branch. There are no run-round facilities on the branch, nor are there any on the Kelbit site, so empty trains are propelled all the way from Ashton-in-Makerfield to Wigan Springs Branch. Class 31 No 31199 is pictured as it sets off in reverse gear down the former Haydock branch with 7E34, the 09.15 'empties' to Lindsey, on 22 February. The siding on the left leads into the Kelbit works, and the stanchions of the electrified West Coast Main Line can be seen just below the horizon.** ***Paul Shannon***

Chapter 5

RAILFREIGHT DISTRIBUTION

Undoubtedly the main railway news story of 1990 was the Speedlink dilemma which first surfaced in the press in the summer. Railfreight Distribution had been in the process of analysing all its freight traffic to try and formulate a portfolio of freight services which would meet the government guidelines for a profitable freight railway. In the process the heavy losses of Speedlink services had become apparent and in the face of these losses the proposal to abolish the Speedlink Network was announced in December. Such changes somewhat overshadowed the more positive developments within the sector during 1990.

FREIGHTLINER AND SWAPBODIES

The majority of Channel Tunnel traffic is likely to be containerised and with this in mind there has been an expansion of the Freightliner services offered to customers during 1990, several of the new traffic flows being illustrated in this chapter. The most noteworthy of these is the new daily 'Swapbody' service from Harwich to Garston. This is a template for Channel Tunnel trains and should eventually run at 90 mph, allowing a 24-hour turn-round for traffic from the south coast to Glasgow.

In addition to the new 'Swapbody' train there have been new services between Cardiff and Holyhead, Coatbridge and Felixstowe and Crewe to Leeds. These new trains cater for two of the areas of Freightliner expansion: deep sea container traffic and Irish traffic. One further new diagram which merits comment is a new Freightliner working to the Isle of Grain to serve the port there, another example of Freightliner competing for the country's expanding container business.

TRAINLOAD

Railfreight Distribution operates both wagonload and trainload services; indeed, many of the country's best-known block trains run under the aegis of Railfreight Distribution. The majority of such services are designated as Chemical and Industrial mineral traffic, and the limestone from Tunstead to Northwich is perhaps one of the best-known such traffic flows. New trains have been introduced during 1990 in this category. Chlorine from Wilton to Langley Green, methanol from Haverton Hill to Barry and chemicals from Immingham to Stalybridge were just some of the commodities to merit new trainload services during the year.

Car traffic also runs as a trainload sub-sector within Railfreight Distribution. As well as additional trains between Swindon and Longbridge there have been new car trains from Longbridge to Harwich carrying export vehicles bound for the continent. This expansion in the trainload arm of Railfreight Distribution is likely to accelerate during 1991 as the end of Speedlink approaches. The history behind Speedlink and the reasons for its demise are examined below.

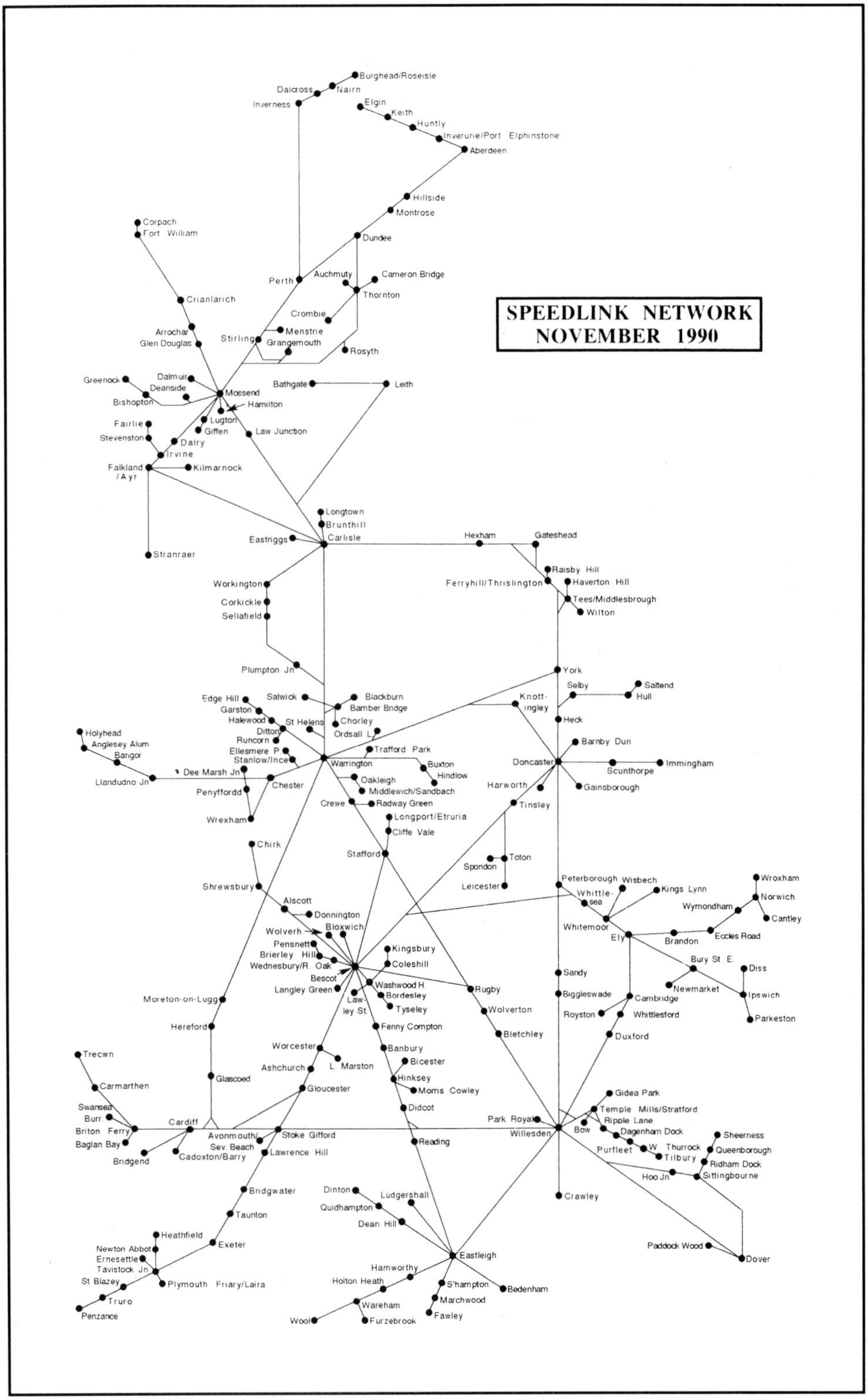

SPEEDLINK NETWORK NOVEMBER 1990

THE SPEEDLINK STORY

When the 1955 Modernisation Plan was published there was no motorway network, and the idea of running a railway without wagonload freight was inconceivable. It was therefore not surprising that one of the most pressing demands was the provision of new marshalling yards to speed up the transit of wagonload traffic. As the modernisation of the network proceeded, so new roads were built, the railway began to run more block freight trains, and it began to dawn on many within British Railways that wagonload freight had a limited future. These concerns were expressed most graphically in the 1963 Reshaping of British Railways report, better known as the Beeching report.

The plans for freight in the 1960s included a new freightliner network with over 50 depots, increased block train working, the introduction of the 'Merry-go-round' (MGR) coal train principle and the contraction of the wagonload and sundries network to just 100 depots. This was radical planning in 1963 when over 5,000 terminals remained open to general goods traffic. The freightliner network had a traumatic birth with just 28 depots finally being authorised. As for the new wagonload network, it never really happened. The National Freight Train Plan of 1966 sought to rationalise trunk freight service operation, but investment and modernisation of wagonload services was not forthcoming.

Under Sir Henry Johnson the consistent attempts to abandon wagonload freight were dramatically reversed with speculation about new rolling-stock, the TOPS computer revolution and the birth of Speedlink. In retrospect this was clearly a mistake as Speedlink services have almost certainly been bottom-line loss-makers since 1973 when the first train was introduced between Bristol and Glasgow. It operated with British Rail-owned VDA vans and OBA open wagons capable of 60 mph running. Stops at Bescot and Warrington provided access to the large industrial conurbations, and departure from Bristol at 18.34 allowed arrival in Glasgow at 05.50 the next morning.

In October 1973 a second service started between March and Edinburgh and by 1975 these two trains had been expanded into a network of services to 19 major cities. Initially loadings were good and it was claimed that 20 per cent of traffic carried was new to rail. The network sought to link population and industrial centres rather than marshalling yards. As the 1970s drew to a close, however, two problems forced a major change in the network. Incompatibility between traditional vacuum-fitted wagons and the new air-braked vehicles often meant that two trains were covering exactly the same route when one train would suffice. As more private companies sought to take advantage of the trains, so the connection of private siding traffic to the trunk workings became more complex. Instead of starting from the British Rail-owned goods depots, trunk trains started from marshalling yards to allow better connections for all their customers. Soon Speedlink was no longer a discreet network but simply a mirror of the pre-1960 wagonload network, albeit air-braked.

Costs spiralled as trip working and marshalling increased. In 1980 the network consisted of trains linking 12 trunk marshalling yards, and there were nearly 20 secondary yards and over 200 freight loading points. The pattern of the 1980s has been a relentless closure of marshalling yards and a gradual reduction of the terminals served by Speedlink. At the start of 1990 just eight main yards remained and of the 200 freight sidings only 100 were used regularly for traditional wagonload traffic. The other sidings all received predominantly bulk sub-sector traffics (coal, cement, oil and steel) or were only serviced by block freight trains (eg Barnby Dunn sand).

The establishment of Railfreight Distribution (RfD) by the combination of Speedlink and Freightliner in October 1988 has provided the platform from which the bottom-line costs of wagonload freight services can be assessed. As suspected, Speedlink is a heavy loss-maker, £30 million being lost on a turnover of £45 million in the last financial year. Faced with this annual financial haemorrhage, RfD began a lengthy review process.

The review itself began by examining every single Speedlink freight flow. The sad findings of this study were that every single freight flow failed to reach profitability. The current Speedlink network, with all its marshalling and tripping costs, requires customers to send goods at least 500 miles and preferably to generate at least ten wagonloads each day. In an island the size of Britain this is simply not possible. Indeed, these demands are equally difficult for trainload traffic, but because of the lack of marshalling and tripping the economics of trainload operation are completely different. It will be remembered that many short-distance coal flows are highly profitable for Railfreight.

The analysis also revealed the surprising information that only 20 per cent of the cost of running Speedlink was taken up by the trunk haul of wagons between major centres; trip working from the major yards accounted for 50 per cent of Speedlink's total costs, and this sum was largely independent of the number of wagons picked up. The marshalling of trip freights into trunk services accounted for the final 30 per cent of Speedlink's direct costs. The reader can therefore see that trainload operation from siding to siding can avoid 80 per cent of the costs incurred by Speedlink. Thus, given suitable locomotive diagrams and freight service planning, trainload operation can be made profitable.

Faced with these findings, RfD had several options. The traditional approach to an ailing railway network is to trim back to a core network which can be profitable, which was the concept behind Beeching and the long-buried Serpell report. This was rendered impossible with Speedlink because of the even geographical distribution of RfD's customers. There is no core network which is intrinsically profitable – the malaise of Speedlink is countrywide.

Another option was to increase the tonnages of traffic handled and thereby move services over the profitability line. This too was fraught with potential difficulty, not least because most Speedlink assets were life-expired and would need replacing during the next decade. Similarly, any increase in traffic would need a duplication of most trunk services because they are fully loaded at the moment. Add to this the prospect of more complex trip working to serve the new business generated by a new marketing drive, and it is not difficult to see why such a plan might not work. In spite of these objections, RfD commissioned a survey of 30,000 companies, all with over 25 employees and all within reasonable proximity of the rail network. Its aim was simple – to assess the potential for generating new Speedlink business. Only ten companies expressed an interest, a 0.03 per cent response rate! There was clearly no market need for a new 'super Speedlink' or 'son of Speedlink'.

A third avenue worth exploring was cost cutting within the current network. A target of a 40 per cent reduction in direct costs was set and the planners redrew a specification of Speedlink compatible with this. The first goal was to improve wagon utilisation by cutting turn-round time from 11 to 9 days, which would save 16 per cent of Speedlink's bill. It is worthy of comment that wagon turn-round time in 1960 was 11.5 days; very little has changed in the wagonload freight arena. Next, train crew productivity had to be increased by planning that each driver would actually drive for 6 hours out of his 8-hour shift; a 6 per cent cost saving. A further 6 per cent could be saved by a 25 per cent cut in marshalling costs and 5 per cent saved by increasing the number of customers served per trip. A 50 per cent increase in the number of annual locomotive hours squeezed from the ageing Class 47s (from 2,000 pa to 3,000+ pa) would give a 4 per cent annual saving as long as maintenance costs did not also rise. Finally, another 3 per cent could be saved by increasing the average number of wagons on a trunk train from 20 to 30. This speculative 40 per cent saving was then applied to every single Speedlink traffic flow – only 15 per cent of flows became profitable. This left 85 per cent of traffic as a loss-maker in spite of a 40 per cent saving for Speedlink. On top of this disappointing finding, the reader will notice how idealistic many of the planners' cost-cutting exercises were.

The final nail in the Speedlink coffin came from the customers themselves. Transport managers all over the country are demanding 'relevance' from the providers of transport. This means that everybody wants goods delivered between 07.00 and 09.00 in the morning and wagons picked up between 16.00 and 18.00. This trip freight method of connecting customers to the main network renders this impossible unless each customer has a dedicated locomotive and crew at his beck and call.

RfD were left with a network which carries 2 per cent of all Railfreight and just 0.12 per cent of all freight in the UK. 81 per cent of journeys were under 300 miles and 63 per cent of trip pick-ups were for fewer than five wagons. Not only was this intrinsically a loss-maker but comparison with road haulage over similar distances revealed the futility of trying to compete for traffic which was never going to benefit the railway without government subsidy. As head of the management team, Tim Hansford had no choice but to advise that the Speedlink network be abolished. The environmental and political wisdom of such a move could not be part of RfD's consideration. Their remit was simply to try and comply with government guidelines demanding a 4 per cent return on all assets by 1993.

The recommendations of RfD that Speedlink should be discontinued were accepted by the British Railways Board on 6 December 1990 and the confirmation of this decision made public on the 12th. From 8 July 1991 there will be no Speedlink network and all traffic running on services denoted SLK in the timetable will be withdrawn. This does not mean that all the traffic currently running on Speedlink will be lost; indeed, it is anticipated that 50 per cent of custom will be retained by the introduction of block train operations.

INDIVIDUAL TRAFFICS ASSESSED

The 380 locations listed in the current RfD timetable as serviced by Speedlink can be broken down according to the traffics they handle. After removal of locations which have already lost their service (eg those handling Redland roof tiles) and sites which do not actually generate or receive traffic (eg marshalling yards), there are just over 300 terminals left in use. Just over 150 of these handle traffic which belongs to one of the other bulk sub-sectors but is delivered by means of Speedlink services, or traffic which already constitutes a profitable bulk load, eg automotives. The function of the final 144 terminals is divided as follows: general use, 52; timber/grain/china clay, 37; BR fuel oil, 18; government stores, 24; and fertilizer, 13.

BULK SECTOR TRAFFIC

Railfreight's bulk sub-sectors will in future take on full responsibility for delivery of their customers' goods. Speedlink has been used over the last two years for a small percentage of Metals, Construction and Petroleum traffic, but the enlargement of the Metals network to encompass places like Sheerness means that this sub-sector is already operating independent of Speedlink. Railfreight Construction has also gradually increased its independent network so that many cement depots like Heaton and Newcastle receive cement on trains resourced by the parent sub-sector. This pattern of operation will increase during 1991 so that distribution from the major cement works can continue without Speedlink.

The Petroleum sub-sector also uses Speedlink to deliver fuel oil to many BR locomotive depots. During the last few months it has introduced new services from Fawley to Ripple Lane and Immingham to Yorkshire which service several locomotive depots *en route*. The possibility of more of these dedicated freight workings means that the majority of fuel oil will continue to be delivered by rail. There are, of course, examples where this might not be possible and more isolated depots like Buxton may suffer the ignominy of road delivery.

RfD AND TRAINLOAD TRAFFIC

Much of the non-freightliner traffic carried by RfD is already in trainloads. The bulk of the chemicals and industrial minerals on the railways form profitable bulk flows. The success of the Wisbech to Deanside service and pet food services to Melton Mowbray is well documented, while the car manufacturing industry generates many siding-to-siding block trains as do other well-known customers like English China Clay with their Fowey Docks operation and the Burngullow to Irvine service.

These varied freight workings form the template for other potential trainload services. The process of consultation with customers is already at an advanced stage and it is anticipated that several well-known Speedlink customers may find a weekly or twice-weekly block load just as attractive as a daily Speedlink trip.

INTERNAL USER TRAFFIC

The establishment of an independent engineers freight network called Civilink gives internal traffic freedom from the Speedlink network. Such traffic was always the poor relation of revenue-earning goods and was often relegated to the most circuitous route between two points. Several busy yards like Doncaster Belmont and Warrington Arpley actually banned such wagons from their sidings in order to try and reduce delays to other Speedlink traffic. This is no longer the case, as dedicated services operate throughout the country to cater for BR's engineering needs. It may be that the carriage of crippled wagons could also be dealt with through this or a similar wagon repairers network as yet to be formalised.

FERTILIZER

This small but important segment of the RfD market is largely carried in block trains from Ince and Elton. The future of wagonload carriage is vulnerable, but there may well be potential to explore expansion of trainload working, particularly once the ownership of the major manufacturing companies is clarified.

TIMBER, GRAIN, CHINA CLAY AND GOVERNMENT STORES

The controversial aspects of the imminent loss of the Speedlink network centre around these four commodities. Customers for the first three are based in rural areas like East Anglia, Devon, Cornwall and Scotland. Indeed, the 37 terminals dedicated to such traffic are almost exclusively in these areas. The future for these traffics looks mixed. RfD has expressed a wish to preserve as much of this environmentally sensitive traffic as possible, but clearly without appropriate government subsidy this may not include all the customers currently benefiting from Railfreight delivery. The Speedlink Review also contains a commitment to continue delivery of government stores, so this aspect of rail operation is largely unaffected by the current rearrangements.

China clay is already mainly delivered in trainloads (see above), and the residual traffic travels mainly from Cornwall to Scotland. Tiger Rail has issued a press statement suggesting that the danger to such traffic may be overstated by critics of the Railfreight Review. Indeed, informed sources suggest that the provision of services sponsored jointly by RfD and Tiger Rail may keep the majority of Anglo-Scottish traffic on the railway.

Grain carriage is more problematic because of the volatile nature of the grain market. It is a credit to RfD that it has already identified a potential East Anglian pick-up system capable of forming a block train to Scotland. Potential problems for Scottish delivery lie in the large distances between terminals in Fife and the Far North. This, coupled with the fact that only nine polybulks can be taken over Druimuachdar summit, limits the options for a profitable trainload operation.

Like grain, timber currently utilises too many different terminals. The annual rail tonnage of 100,000 tonnes is collected from multiple loading points, nearly all handling well below 10,000 tonnes per annum. The major customers for timber at Chirk and Shotton utilise 80 per cent of this wood, so there is clearly scope for trainload planning if the area of supply can be narrowed down. All three commodities could be rail carried if a profitable train plan can be formulated.

GENERAL GOODS AND THE CONTINENT

If all the above traffics are deducted from the Speedlink terminals list there are just 52 depots left. Many of these handle well-known products like cider or beer. The majority also handle goods from the continent, brought into the country via the Dover train ferry. Clearly this is traffic that should be encouraged rather than withdrawn. RfD anticipates that nearly 85 per cent of continental goods traffic will be retained by the establishment of train services based at Dover and servicing 40-50 depots around the country. Such workings will of course form a template for Channel Tunnel freight services due to start in 1993, and will form a valuable precursor to the future RfD train plan. The likely effect of the abolition of the Speedlink network on general terminals will be softened by the intention to retain those depots where sufficient continental traffic is present.

PRIVATISATION

RfD has declared its willingness to talk to anyone interested in running Speedlink services privately. Indeed, such negotiations are already under way with several wagon manufacturers. The simplistic idea that Speedlink can be privatised in its entirety is clearly not a viable one. No potential investor is going to sink millions of pounds into a business which lost £30 million last year and which has just been examined by independent consultants, whose findings are discussed above. While only 0.03 per cent of companies questioned show an interest in the product and even with a 40 per cent cost cutting exercise 85 per cent of trains still make a loss, wholesale privatisation is just a dream.

Railfreight Distribution has, then, decided to withdraw from wagonload freight handling and in doing so abolish the Speedlink network set up 17 years ago to try and stem the flow of freight from road to rail. The loss of traffic has continued and Speedlink has been unable to respond. However, investment in relevant freight services will continue and expansion of the continental market is anticipated. Railfreight has acted in order to comply with government guidelines, and the financial wisdom of its plans is undoubted; whether the loss of Speedlink is a major environmental or political mistake is another question. Envious comparisons with the continent are fallacious as a recent report on the crisis in SNCF freight handling has revealed. Railfreight has undertaken a review which is unlikely ever to be repeated, and has certainly provided no other option than the abolition of Speedlink.

SERVICE CHANGES

Wirksworth-Kings Lynn: **The 1989-90 sugar beet season was the last in which limestone was hauled by rail from Wirksworth quarry in Derbyshire to Kings Lynn. The ageing HTV hoppers used for the traffic were in need of renewal and investment could not be justified for just four months' use each year. Towards the end of the season, on 12 January, Class 47 No 47119 approaches March with 8M06, the returning empties from Kings Lynn to Derby St Mary' s yard.** ***Brian Morrison***

Haverton Hill-Barry: **During 1990 a new flow of methanol started to run from Haverton Hill on Teeside to Barry (Dow Chemicals). On 7 May the train is pictured passing Bromsgrove behind Class 47 No 47363. The train is coded 6Z15, and is running as a special as it has not yet found a permanent place in the working timetable.** ***Don Gatehouse***

Longbridge-Harwich: Class 47 No 47216 approaches Caledonian Road station on the AC electrified tracks of the North London line with 4Z51, the 06.10 Longbridge to Harwich 'special', on 1 November. The train was running five days a week during October, November and December of 1990, loaded with Rover cars of various shapes and sizes for export to Europe. *Paul Shannon*

Melton Mowbray-Dover: Railfreight Distribution is keen to exploit the potential of trainload freight between Britain and mainland Europe in advance of the opening of the Channel Tunnel. One new service which has run as required during 1990 brings pet food via the Dover Dunkerque train ferry to the Pedigree private siding at Melton Mowbray. This scene at the west end of Melton Mowbray station shows Class 47 No 47018 backing out on to the main line with 15 empty Cargowaggon bogie vans for Dover on 17 August. Traction resources for services such as this are taken from the national Railfreight Distribution pool; only two days previously No 47018 had been engaged on Speedlink trip workings in the Eastleigh area. *Paul Shannon*

Stanlow-Immingham: A new block train takes propylene from the Sheel refinery at Stanlow to the Conoco refinery at Immingham (Humber). Passing through the Cheshire countryside near Moore on 2 May is a unidentified Class 47 from the Stanlow petroleum fleet with 6E05, the 17.25 Stanlow to Immingham working. The train has been scheduled to run each weekday and is routed via Manchester Victoria and the Calder Valley line. *Paul Shannon*

Willesden (Distillers): **Distillers MG, one of the UK's leading suppliers of industrial gases, opened its new distribution and technical centre at Willesden on 19 November. The centre, covering 4.2 acres, is rail connected and has been built on the site of the former Minilink terminal; the site was chosen because of its proximity to Brent marshalling yard just one mile to the north. This was the centre for Speedlink operation in the London area and Distillers saw the potential of using Speedlink to distribute its products throughout the UK (RfD announced the closure of the Speedlink network just two weeks after this picture was taken at the opening of the rail terminal). Class 47 No 47214 was named *Distillers MG* by Ian Brown, the Director of RfD, and Mr K. A. Wokalek, Commercial Counsellor from the German embassy. *Steve Knight***

British International Freight Association: **Class 37 No 37194 leaves Glasgow Central on 9 September after being named *British International Freight Association.* The naming was carried out by Lady Denholm in the presence of Ian Brown from RfD and Cyril Bleasdale, Director of Scotrail. Glasgow was chosen as the location for the ceremony because of RfD's commitment to compete for international traffic to Scotland. *Tom Noble***

Middleton Towers-Worksop: **British Industrial Sand (BIS) ran wagons from Middleton Towers near Kings Lynn to customers at Monk Bretton and Kirk Sandall in Yorkshire until a new destination was added in 1990, Worksop. The train of sand usually travels from Kings Lynn to Doncaster behind two Class 31 locomotives. From Doncaster trips are made to all three terminals before the assembled empties return south from Doncaster as a block load. On 13 November Class 31 No 31184 stands in the Worksop terminal at the head of six empty wagons which will return to Doncaster that afternoon.** ***Michael Rhodes***

Swindon-Longbridge: **Such is the tonnage of traffic between the British Leyland car plants at Swindon and Longbridge that there are now two daily freight services between the works. On 4 April, Class 47 No 47297 is seen banking 6M03, the 12.10 Swindon to Longbridge service, up the Lickey incline. The train is hauled by Class 47 No 47335 and is the only regular daylight working to need assistance up the incline.** ***Don Gatehouse***

Tinsley Depot: **Motive power for the national Speedlink Network is maintained at Tinsley depot near Sheffield. The engine shed here was built in 1965 as part of the large hump marshalling yard at Tinsley. Recently the yard has been relegated to a terminal yard for some Metals sector traffic, but the depot continues to be busy with the Speedlink fleet. On 4 July Class 37 No 37032 stands amongst other Speedlink locomotives in the maintenance shed. The future of the depot must be under review, however, as there will be no Speedlink network from July 1991.** ***Michael Rhodes***

SCOTTISH SPEEDLINKS

Bishopton trip: The 6T10 trip from Mossend services the British Aerospace plant at Bishopton. Here, on 20 August, Class 37 No 37154 crosses over on to the up main line at Bishopton station with the afternoon trip back to Mossend yard. The Speedlink service will connect with Anglo-Scottish trunk services at Mossend; its consist of VEA vans carries government stores. *Tom Noble*

Greenock-United Molasses: **Perhaps Scotland's best-known Speedlink traffic is molasses from Greenock to Menstie in Clackmannan. The United Molasses terminal at Greenock typifies a small Speedlink railhead with the capacity to handle just six molasses tanks at a time. On 26 November Class 37 No 37019 awaits departure with two tanks, each of which takes 20 minutes to fill. The train will run as 6T19 to Mossend yard from where it will depart the next morning to Menstie.** ***Tom Noble***

Grangemouth-Mossend: **Class 20 Nos 20066 and 20185 pass Greenfoot, south of Cumbernauld, with 6N06, the Grangemouth to Mossend trip working on 31 August. Four different types of tank wagon are included in the formation: liquid petroleum gas (loaded), molasses (empty), cement (empty) and diesel fuel (loaded). Both locomotives were members of pool FPAE, belonging to Railfreight Petroleum although at work on a RfD service. The molasses tanks will return to Greenock (see above).** ***Paul Shannon***

Mossend-Aberdeen (1): **Aberdeen is relatively busy with freight traffic, much of it connected with the paper and oil industries. In summer 1990 there were two daily Speedlink workings in each direction between Mossend and Aberdeen. The daytime southbound working, 6M64, the 12.32 Aberdeen-Mossend-Willesden, approaches Arbroath on 29 August with a healthy and varied load, including ferry vans for the continent, empty china clay tanks for Cornwall, timber for Shotton, containers from Elgin to Holyhead, and an empty starch wagon for Manchester. Additionally on this occasion the train had called at Montrose to pick up empty lime wagons for Ferryhill, following the cancellation of 6E89, the Aberdeen to Tees service which normally conveys this traffic.** ***Paul Shannon***

Mossend-Aberdeen (2): **Speedlink traffic reaches the 'Granite City' by way of services from Mossend yard. On 10 April, 6A19, the 07.50 Mossend to Aberdeen Speedlink service, roars past Lunan Bay behind Class 37 No 37262** ***Douneray.*** **The locomotive is painted in all-over grey and belongs to the Departmental sub-sector, but on this occasion finds employment hauling china clay bound for the paper mill at Port Elphinstone and cement to Aberdeen.** ***Les Nixon***

Elgin (1): **Class 37 No 37113 shunts CO_2 tanks in Elgin yard on 26 September. The yard at Elgin is serviced by two daily trips from Aberdeen Guild Street where traffic connects with the main Speedlink network. Traffic is usually light and made up of timber, occasional government stores, whisky and CO_2. In times of international conflict, such as the Gulf Crisis, traffic is greatly augmented by government stores to and from the air bases in the north of Scotland (see below).** ***Steve Turner***

Elgin (2): **An extra Speedlink service from Elgin to Aberdeen is seen reversing out of the yard at Elgin on 26 September. Class 47 No 47321 propels an assortment of Railfreight-owned VDA and VGA vans hurriedly assembled to deal with the demands of the armed forces responding to Iraq's invasion of Kuwait. The retention of large numbers of general-purpose freight vehicles is difficult to justify in an age of 'bottom-line' accounting, and it is unlikely that the RfD of the future will own large numbers of vehicles for such occasional uses.** ***Steve Turner***

Burghead: **Grain from East Anglia has traditionally been carried by rail to the distilleries of northern Scotland. Class 37 No 37209 winds out of the distillery at Burghead with the 6H11 trip to Inverness, from where the empty grain wagons will return to Mossend and then East Anglia. The preservation of such traffic after July 1991 will depend upon rationalising traffic flows so that they run only between the major suppliers and users of grain.** ***Steve Turner***

SPEEDLINK CASE STUDY 1 : THORNTON, FIFE

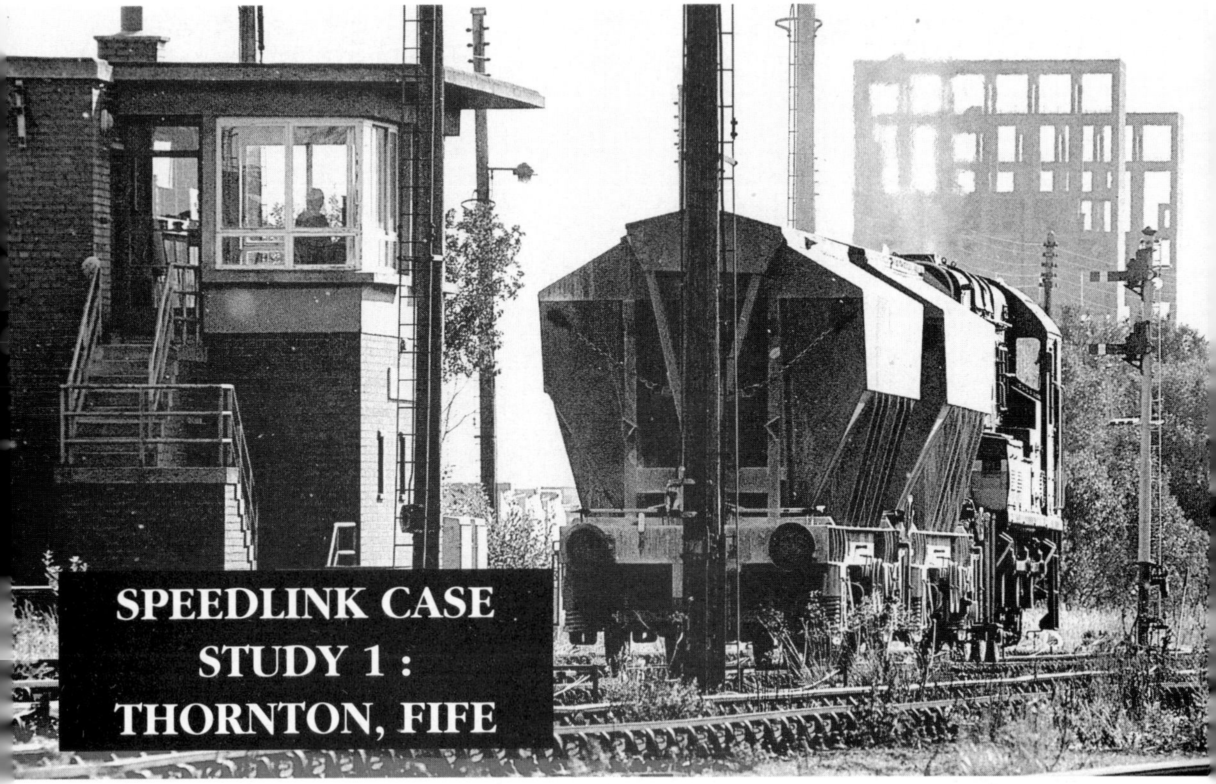

Thornton Yard (1): **Class 08 No 08761 shunts at the east end of Thornton yard on 17 August. The two china clay wagons are bound for Auchmuty and have arrived at Thornton on 6G43, the morning Speedlink from Mossend. The yard pilot leaves Thornton at about 10.00 for a morning trip to Auchmuty, returning after 12.00. The concrete shell behind the train is the remains of Rothes mine which was built in the 1950s but flooded before opening and has never produced any coal.** ***Michael Rhodes***

Auchmuty branch: **The delightful Auchmuty branch continued to receive regular traffic in 1990 in the form of china clay from South West England to the Tullis Russell paper mill at Auchmuty. There are no run-round facilities at the end of the branch, so a brake van is provided to allow the locomotive to propel its empties back from Auchmuty to Markinch. Class 08 No 08761 was in charge of the morning trip from Thornton yard on 28 August. The train is pictured as it starts to back out on to the main line at Markinch on the last leg of its return to Thornton.** ***Paul Shannon***

Crombie branch: **On 17 August, Class 20 Nos 20118 and 20137 burst out of the densly wooded country on the north bank of the Firth of Forth at Crombie. The 6G07 trip from Thornton yard brings two VEA vans to the RNAD depot and will depart later with five vehicles bound for the south. The activity on this delightful rural branch line was greatly increased in August because of the invasion of Kuwait by Saddam Hussein.** ***Michael Rhodes***

Elbowend Junction: **Class 20 Nos 20211 and 20198 are captured winding up the steeply graded line from Crombie to Elbowend Junction on 27 April. Two VDA vans used as barrier vehicles are all that needs to be returned to Mossend on this date. The two engines will visit Rosyth and Cameron Bridge as required before returning to Mossend yard in the afternoon.** ***David Hamilton***

Dumfermline Townhill, 1980: **Freight traffic from Rosyth and Crombie was marshalled at Townhill yard in Dumfermline ten years ago. On 29 June 1980 Class 25 No 25011 passes Townhill with 9B41, the morning trip from Rosyth to Townhill yard. After reversing into the yard the engine made a second trip to Crombie with government stores.** ***Michael Rhodes***

Dumfermline Townhill, 1990: **On 17 August 1990, Class 20 Nos 20137 and 20118 have just arrived at Dumfermline Townhill with the 6G07 trip from Crombie, pictured in almost the same position as Class 25 No 25011 in the picture above. Gone are the semaphore signals, the marshalling yard and associated locomotive depot at Townhill. The Class 20s will run round their train here and continue to Rosyth Dockyard before returning to Thornton yard. Thus the 1990 trip freight will visit Crombie and Rosyth in the reverse order as compared to its 1980 forerunner.** ***Michael Rhodes***

Rosyth Dockyard: The second duty of the morning for 6G07 on 17 August 1990 was to pick up four ODA wagons at Rosyth to add to the five VEA vehicles from Crombie. Class 20 Nos 20137 and 20118 are seen winding out of Rosyth on the final leg of their morning trip from Thornton. The platform used for the daily workingmen's train to Rosyth can just be seen above the ODA wagons. *Michael Rhodes*

Rosyth Dockyard signal cabin: Train movements within Rosyth Dockyard are controlled by this North British signal box. Built in 1917, the box is due to be demolished in 1991 with the installation of handpoints and the removal of signals. It must be hard to justify its retention when there is often only one train each day to the dockyard. *Michael Rhodes*

Cameron Bridge: Class 37 No 37262 departs from Cameron Bridge with two carbon dioxide tanks for Coleshill on 27 August. The wagons will be processed at Thornton from where they will travel on to Mossend on the 6G43 working. From there they will travel south to Bescot, reaching Coleshill the next morning. Two further carbon dioxide tanks which await loading can be seen beyond the footbridge. *Paul Shannon*

Thornton Yard (2): This general view of Thornton yard taken on 17 August shows a wide variety of freight traffic and motive power. Class 26 No 26026 stands on the right, having just returned from Perth with a ballast working. Also on the depot are Class 20 Nos 20124 and 20138 which worked the 7L22 Thornton to Laurencekirk freight earlier in the day. In the distance the remains of the hump marshalling yard are visible. Yard pilot Class 08 No 08761 shunts HAA hoppers bound for Westfield whilst Class 20 Nos 20137 and 20118 have just arrived with 6G07, the morning trip from Crombie and Rosyth. These engines will visit Cameron Bridge in the afternoon before returning to Mossend yard. To the right of these Class 20s is a single empty PBA china clay hopper which was brought from Auchmuty paper mill by the yard pilot. *Michael Rhodes*

NORTH EAST SPEEDLINK

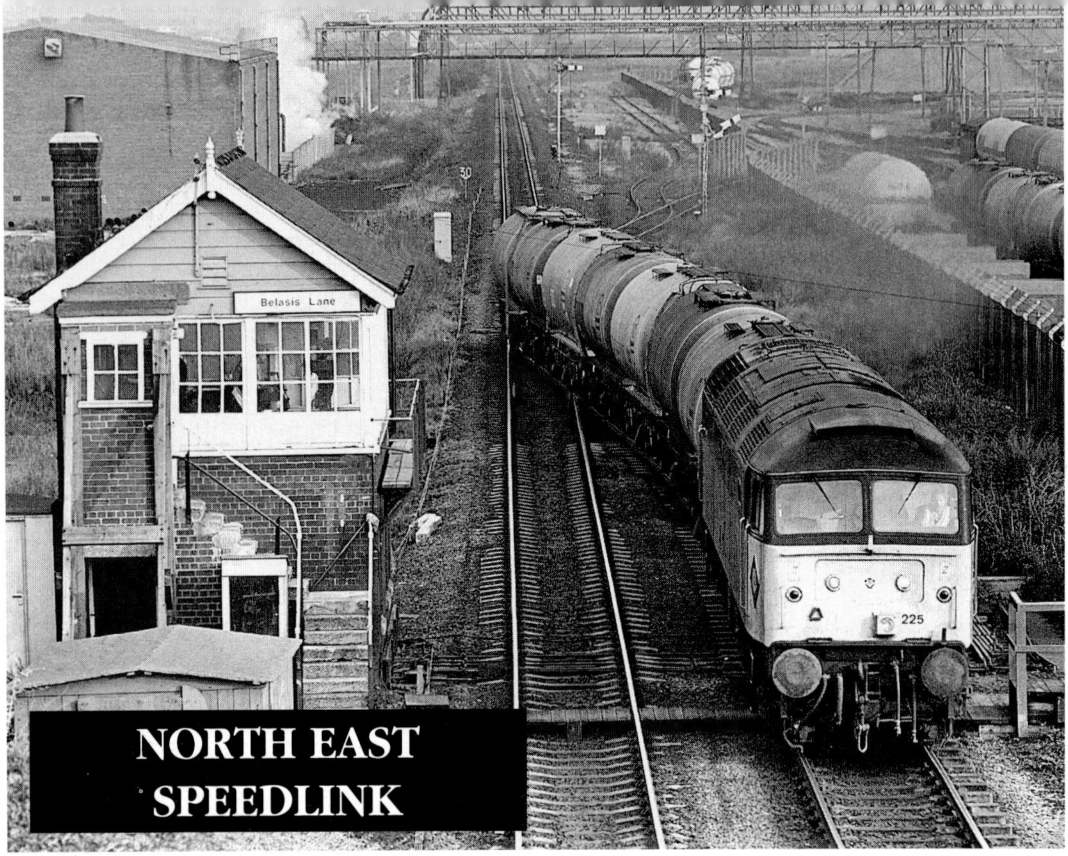

Haverton Hill: One of the major Speedlink customers in the North East is ICI. The announcement that the fertilizer plant at Haverton Hill is to close in 1991 has several implications for rail traffic. Ammonia traffic from the site to Leith and Severnside will eventually cease, but a weekly train to Stevenston is likely to continue. Leith is probably to be supplied from a plant at Saltend near Hull and there may be new trains between Severnside and Ince Marsh.

The ICI Carbon Dioxide plant at Billingham is also being sold, and clearly rail delivery of the gas will depend upon the ultimate purchaser. Distillers currently use most of the rail output from the plant and if they were to purchase the facility then rail carriage would probably continue.

ICI no longer manufactures urea at Billingham, so this traffic to Duxford no longer runs. Another important flow, potash from Boulby to Middlesbrough goods depot, will also cease when fertilizer production finishes. Hydrogen cyanide will continue to be carried by rail, as will petroleum and methanol. The last of these commodities, methanol, is seen winding out of the Haverton Hill complex in September 1990. Class 47 No 47225 heads past Belasis Lane signal cabin with 6L81, the afternoon Speedlink to March Whitemoor yard. *Michael Rhodes*

Thornaby: Class 47 No 47371 storms out of Tees Yard on 22 May at the head of the 6M74 service to Willesden Brent yard. The first half of the train consists of steel girders, made at BS Skinningrove and bound for Italy. Even though they are Metals sub-sector traffic, they are carried by Speedlink to Dover via Willesden. Such goods will have to find another way to the train ferry when Speedlink is abolished in July 1991. *Michael Rhodes*

Gateshead Freight Depot: **Both the labour-intensive nature of Speedlink and its potential pitfalls are illustrated in this view, taken at Gateshead Freight Depot on 14 June. Class 47 No 47217 has just arrived with 6N73, the daily Tees to Hexham Speedlink trip service. Class 08 No 08544 has duly drawn on to the rear of the train to remove traffic destined for the Newcastle area. The confused look on the faces of the staff at Gateshead is because half of the train is missing!** ***Michael Rhodes***

Middlesbrough: **The Stockton Haulage freight depot in Middlesbrough is home for two 0-4-0 industrial locomotives. On 25 April, No 2 is seen on the left repositioning vans from Taunton which have brought cider to the North East. The carriage of soft drinks is an area which is particularly likely to transfer to road transport when Speedlink disappears.** ***Michael Rhodes***

Boulby-Middlesbrough: **Potash from the mine at Boulby travels by rail to Middlesbrough Goods Depot from where it is carried by road to ICI Billingham. This traffic is due to cease in 1991. On 22 May Class 20 Nos 20138 and 20156 pause in the old down departure sidings at Tees Yard with the 6P66 diagram. This working runs between Boulby and either Middlesbrough Goods or Tees Dock. On this occasion the train is for Middlesbrough Goods and therefore has to run round in Tees Yard before returning east for just under one mile to Middlesbrough.** ***Michael Rhodes***

NORTH WESTERN SPEEDLINK TRIPS

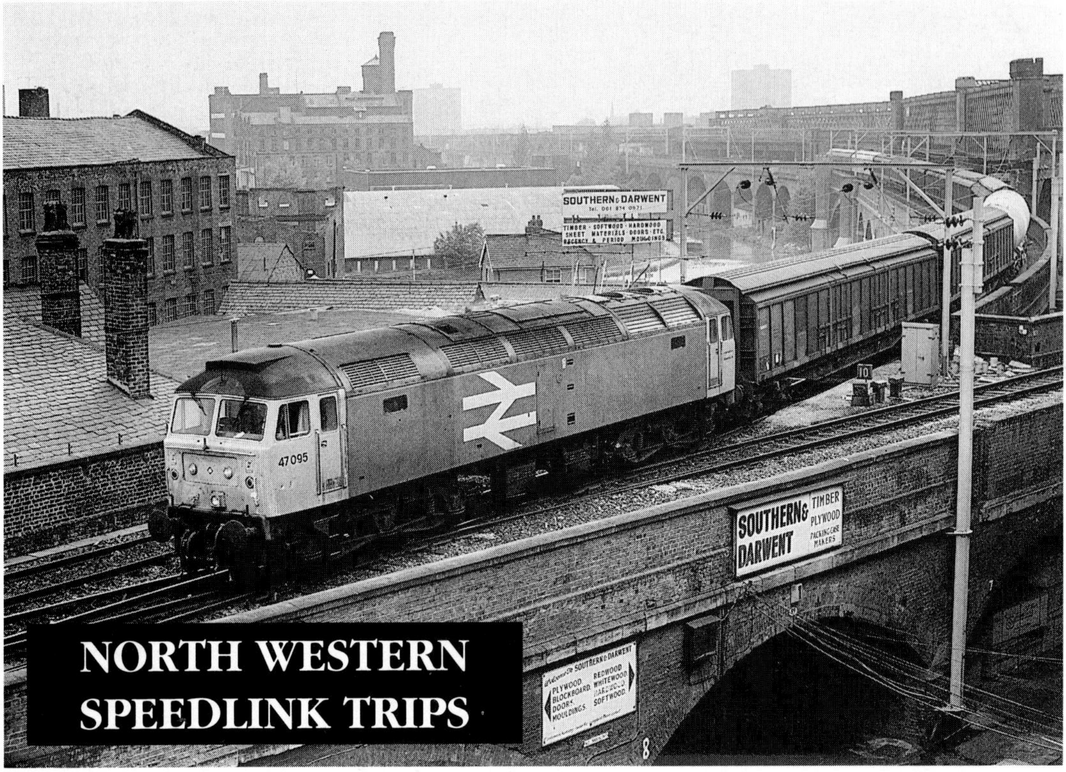

Trafford Park-Warrington: Photographed from the G-MEX car park which will soon revert to railway use when Metrolink comes on stream, Class 47 No 47095 approaches Deansgate with 6F32, the 18.09 Trafford Park to Warrington Arpley yard Speedlink trip working. Apart from ferry vans for Dover, the formation includes three Cerestar starch tanks from Trafford Park industrial estate to Thornton and Aberdeen. From October 1990 all outward Speedlink traffic from Trafford Park was transferred to the morning departure to Warrington, enabling the withdrawal of 6F32. *Paul Shannon*

Salwick: The Blackburn area Speedlink trip sometimes serves Salwick, and when there are wagons to be removed from Salwick the return leg of the trip has to proceed to Kirkham or Poulton in order to run round. On 3 September Railfreight Distribution-liveried Class 47 No 47283 *Johnnie Walker* departs from Salwick with two empty chemical tanks from Salwick and one empty ferry van from Bamber Bridge. *Paul Shannon*

Bamber Bridge: A new rail-connected warehouse was opened at Bamber Bridge in early 1990, to handle general merchandise travelling between the North West and mainland Europe. Class 31 No 31132 negotiates the new trackwork at Bamber Bridge on 12 April as it departs with what was reported to be the first outward consignment of goods from the depot. Visible top left is the East Lancashire line, with Bamber Bridge station roughly half-way across the picture. *Paul Shannon*

Blackburn: One of the most interesting Speedlink operations in the North West in summer 1990 was the sequence of trip workings which served terminals at Blackburn (Fogartys), Bamber Bridge (W. H. Bowker), Chorley (Royal Ordnance) and Salwick (British Nuclear Fuels). Normally one locomotive would set out from Warrington around 09.00 and carry traffic to and from each terminal as required before returning to Warrington in the early afternoon. The same locomotive was also booked to work early morning and late evening trips to Ordsall Lane in Manchester, ensuring excellent resource utilisation. On 6 April Class 47 No 47142 *The Sapper* has just arrived at Blackburn yard with 6N73, the 08.47 trip from Warrington, and 20 minutes have been used to shunt the train. Starting on the left we have: one engineer's wagon for Blackburn station; five VEA vans for Chorley; one Transferry curtain-sided van for Bamber Bridge; and one bogie ferry van for Fogartys depot at Blackburn. Whilst shunting has been going on, Class 37 No 37058 has arrived with 6M58, the 04.30 Lackenby to Blackburn block steel train. *Paul Shannon*

SPEEDLINK CASE STUDY 2 : SOUTHAMPTON AREA

On Tuesday 14 August we followed the movements of the daily Speedlink service from Eastleigh to Dorset. Distribution-liveried Class 47 No 47144 is in charge.

Southampton down yard: **Class 47 No 47144 arrives at Southampton down yard with 6W51, the 08.03 Speedlink trip from Eastleigh to Furzebrook. The only traffic on offer from Eastleigh is one IPA van carrying fertilizer from Immingham to Holton Heath, but the call at the yard is being made to collect two wagonloads of steel for Holton Heath which had been detached here from the overnight Cardiff to Hamworthy steel train.** ***Paul Shannon***

Furzebrook: **The train left Southampton with the van and two steel carriers for Holton Heath but did not call at Holton Heath on the outward journey because the siding is on the up side of the line. The three wagons were taken first down the branch to Furzebrook, where two wagons loaded with china clay were awaiting dispatch. The photograph shows No 47144 about to attach the china clay wagons to the Holton Heath traffic.** ***Paul Shannon***

Worgret: **Upon departure from Furzebrook the train is recoded 6W56, the 13.23 Furzebrook to Eastleigh. The location here is just south of Worgret Junction on the Furzebrook branch, a stretch of line that could soon see the return of passenger trains if the efforts of the Swanage Steam Railway are successful.** ***Paul Shannon***

EASTLEIGH YARD TRIPS

Eastleigh Yard: **Eastleigh yard is the main gathering point for wagonload freight traffic in the Southampton area, with regular trip workings in the summer of 1990 to Dinton, Quidhampton, Furzebrook (see opposite), Holton Heath, Wool, Ludgershall, Marchwood, Fawley and Bedenham. Several of these services are illustrated here. Trunk Speedlink services depart to Willesden, Whitemoor and Tees. On 13 August Class 47 No 47289 departs from Eastleigh with 6L94, the 20.34 to March Whitemoor, which will call at Willesden and Cambridge *en route.* The first three wagons carry fuel oil from Fawley to the Esso private siding at Cambridge Coldhams Lane, this being one of the few petroleum flows, other than those to British Rail depots still to travel by Speedlink. *Paul Shannon***

Marchwood: **The Fawley branch is the last outpost of semaphore signalling on the South Western section of the Southern Region. Class 47 No 47205 restarts briskly after pausing at Marchwood level crossing with 6Z50, an additional 14.30 Marchwood to Eastleigh trip working, on 15 August. *Paul Shannon***

Bedenham: **The Bedenham branch was scheduled to receive a train three times weekly in the summer of 1990. On 15 August, Class 47 No 47249 arrives at Bedenham with one VDA van, one VEA van and two OAA open wagons, forming 6T57, the 10.34 departure from Eastleigh.** ***Paul Shannon***

Southampton Docks: **This rare view taken on 11 October shows Class 47 No 47289 leaving Southampton Docks with 6Z46, the 14.20 to Eastleigh yard. The train conveys empty liquid nitrogen tanks from Berth 105 in the dock area.** ***Brian Denton***

Ludgershall branch: **On 21 November the 6Y06 Speedlink trip from Ludgershall to Eastleigh is seen as it accelerates away from Andover. Class 47 No 47188 heads a typical load for the train which includes six VEA vans and a PFB flat wagon designed to carry road vehicles.** ***Steve Miller***

SPEEDLINK CASE STUDY 3 : OXFORD

Private sidings at Didcot, Morris Cowley and Bicester were served in 1990 by a daily out-and-back service from Bescot yard. The four photographs in this section illustrate the movements of this service on 18 April 1990. Traction is provided throughout by Class 47 No 47285.

Oxford Hinksey Yard (above): **After working from Bescot to Didcot in the early hours of the morning, No 47285 has just run round in Hinksey yard, Oxford, with the second leg of its duties, 6A33, the 08.45 Didcot to Morris Cowley. The formation includes empty vans and coal for Bicester and paper loaded on Freightliner flats from Corpach to Morris Cowley.** ***Paul Shannon***

Morris Cowley: **This is the scene at Morris Cowley half an hour later. The wagons for the freight terminal here have been detached and the train is now ready to depart with vans and coal hoppers for Bicester, running as 6A49, the 11.58 Morris Cowley to Bicester.** ***Paul Shannon***

Bicester: **After a smooth passage through Oxford station and along the single-track section from Oxford North Junction, the train is seen approaching Langford Lane crossing, just short of Bicester.** ***Paul Shannon***

Oxford Banbury Road: **No 47285 passes the ARC roadstone depot at Oxford Banbury Road with 6A50, the 13.56 Bicester to Morris Cowley, conveying empty HEA hoppers from Bicester plus VDA and VBA vans. At Morris Cowley some more empty wagons will be attached and the train will return to the Midlands as 6M93, the 16.20 Morris Cowley to Bescot Speedlink feeder service.** ***Paul Shannon***

CORNISH CHINA CLAY

Heathfield-Tavistock Junction: **Class 47 No 47355 clears the tunnel at Dainton summit with two Tiger Leasing PBA hoppers in tow. The train is the daily Heathfield to Tavistock Junction Speedlink trip and conveys 'ball clay' which will be attached to northbound Speedlink services later in the day. The use of wagons leased from the Tiger Rail Company is perhaps one of the reasons why Tiger Rail has expressed an interest in running china clay services privately after the end of Speedlink in July 1991.** ***Les Nixon***

Exeter-Dover: **Much of the railborne china clay travels in block loads such as the thrice-weekly service from Exeter to Dover. 6O37, the 12.40 Exeter Riverside to Dover is seen at Cole on 8 August behind Class 50 No 50044. The train carries china clay for the export market.** ***Brian Denton***

Coatbridge-Tavistock Junction: **The 6V92 trunk Speedlink service from Coatbridge to Tavistock Junction is seen passing Teignmouth behind Class 47 No 47213 on 9 May. Clearly such a small payload is not cost-effective for a 2,750 hp locomotive capable of 95 mph.** ***Steve Miller***

Tavistock Junction-Gloucester: **Another short Cornish Speedlink is seen passing Totnes on 21 February behind Class 47 No 47258. 6B97, the Tavistock Junction to Gloucester service, conveys a variety of traffic including bagged china clay in a ferry van.** ***Steve Miller***

TRUNK SPEEDLINKS

Willesden-Dover: **Class 47 No 47337 snakes across the complex trackwork at Factory Junction, just north of Wandsworth Road station, with 6O98, the 13.19 Willesden to Dover Speedlink service on 31 July. Behind the 12 ferry vans (just out of view in this picture) are three empty cement tanks on their way back from Aberdeen to Swanscombe; these will be detached at Hoo Junction. 6O98 was demoted to an 'as required' special working from the October timetable change.** ***Paul Shannon***

Bristol-Coatbridge: **The first joint wagonload/Freightliner service to result from the merger of Speedlink and Freightliner in October 1988 was a daily working carrying container traffic between Bristol and Coatbridge. Initially the northbound Freightliner traffic fed into the Speedlink Network at Gloucester, but this was altered to East Usk in January 1990 and then to Hereford in May, the latter being a consequence of Speedlink's withdrawal from East Usk. The 'mixed' train in its latest guise, 6S74, the 19.54 Hereford to Coatbridge, approaches Dinmore tunnel on 9 July, hauled by Distribution-liveried Class 47 No 47079. Accompanying the two Freightliner wagons are an assortment of vans carrying cider from Hereford to depots in northern England and Scotland.** ***Paul Shannon***

Workington-Willesden: **Class 47 No 47245 departs from Corkickle with 6A40, the 16.40 Workington to Willesden Speedlink on 24 July. The train comprises six PCA tanks with sodium tripolyphosphate from Corkickle to West Thurrock, two departmental bogie bolster wagons (one YYA and one YLA) with rails from Workington steel works, and ten TTA caustic soda tanks returning empty from Corkickle to Runcorn.** ***Paul Shannon***

Warrington-Mossend: **InterCity-liveried Class 87 No 87035** ***Robert Burns*** **approaches Winwick Junction on 30 May with 6S80, the 19.54 Warrington to Mossend trunk Speedlink train. The load is mostly chemical tanks from Cheshire to south-west Scotland, but at the rear is a solitary OTA timber wagon from Shotton. The booked traction for 6S80 at this time was a single Class 86/4 locomotive, but members of Classes 85, 87 and 90 were putting in regular appearances.** ***Paul Shannon***

Tyne Yard: **Two long-distance Railfreight Distribution services at Tyne Yard in 17 July. On the left, Class 47 No 47321 leaves the yard with 6S63, the Tees to Aberdeen trainload service. Since 1989 this has run under the Chemicals and Industrial Minerals sub-sector aegis and carries exclusively agricultural lime from Thrislington, Co Durham, to distributors in Montrose and Laurencekirk. On the right, Class 47 No 47378 heads 6S66, the Tees to Ayr Speedlink service. It mostly consists of chemical and cement traffic; the cement is from Eastgate to Carlisle London Road. Although Speedlink pulled out of Tyne Yard in November 1989, long-haul services continued to pause there during 1990 for remanning only.** ***Michael Rhodes***

Baglan Bay-Doncaster: **On 8 August Class 47 No 47245 storms past East Usk with 6E64, the 14.30 Baglan Bay to Doncaster Speedlink. The train carries eight TTA chemical tanks from Baglan Bay to Saltend near Hull, and two newly repaired wagons from Port Tennant works near Swansea. Although designated a Speedlink working, it is clear that the train is effectively a block working of chemical traffic and will no doubt revert to such after July 1991.** ***Michael Rhodes***

Mossend-Stoke Gifford: **7V93, the 07.55 Mossend to Stoke Gifford has been a permanent fixture in the Speedlink timetable for many years. The destination of the train used to be Severn Tunnel Junction until the closure of that yard in November 1987, but the traffic on the train has remained largely unaltered over the last ten years. Class 85 No 85037 hauls a typical consist past Abingdon on 20 June. Included in the load are two empty china clay wagons returning to Cornwall and several VGA vans returning to Taunton Cider.** ***Michael Rhodes***

TRIP FREIGHT MISCELLANY

Duxford-March: **A firm which has invested heavily in rail facilities for wagonload traffic is Ciba Geigy at Duxford. Private sidings were installed on the site in 1980 and commodities handled in recent years have included urea from Kings Lynn, methanol from Purfleet and Haverton Hill, china clay from Cornwall and resin to Stranraer (for Ireland) and Hexham. Smartly painted Railfreight Distribution Class 47 No 47309 departs from Duxford on 7 August 1990 with the 12.38 trip working to March Whitemoor yard. The train consists of various tank wagons carrying resin for Hexham and empty urea hoppers for Kings Lynn.** ***Paul Shannon***

Carmarthen: **A daily Speedlink working continued to penetrate into West Wales in 1990, serving two freight terminals at Carmarthen and one at Trecwn. The outward working, 6B54, the 05.15 Gloucester to Trecwn, backs out of Carmarthen on 13 July behind Class 37 No 37015. The train engine has just berthed two ferry vans in the adjacent siding which contain fertilizer from Immingham. The train now comprises just one VEA van with two barrier wagons.** ***Paul Shannon***

Carmarthen-Bescot: 6M83, the 17.10 Carmarthen to Bescot Speedlink feeder service, storms past Briton Ferry on 10 September. Class 37 No 37358 has just one VEA of revenue-earning traffic from RNAD Trecwn and two barrier wagons (VDA and VGA). The train is unusual in that it is formed of one of each of the most common Railfreight-owned covered wagons. *Geoff Dowling*

Bow: The former Midland Railway goods yard at Bow continued to receive wagonloads of coal, construction blocks and carbon dioxide during 1990, in addition to trainloads of granite from Croft. The wagonload traffic was tripped from Temple Mills yard early in the morning, with empties returning in the afternoon. Class 08 No 08709 shunts one HEA hopper with industrial coal from Gedling plus several tanks of carbon dioxide shortly after arriving at Bow on 2 August. *Paul Shannon*

Hereford: Having been a thriving centre for steel and timber traffic not so long ago, Hereford yard had become rather quiet by the summer of 1990, with the Bulmers cider company providing most of Railfreight's carryings. The resident shunter, Class 09 No 09015, pulls four vanloads of cider out of the yard on 9 July, ready to be attached to the evening departure for Coatbridge. *Paul Shannon*

Toton-Leicester: One problem facing Railfreight in 1990 was what to do with the considerable tonnages of fuel oil which need transporting to BR depots and fuelling points in all corners of the system. The nature of the problem is illustrated by this photograph of Class 37 No 37217 on a Toton to Leicester Speedlink trip working on 23 April. A single BR fuel tank is the only traffic on the train. *Paul Shannon*

Donnington: Class 47 No 47238 draws to a halt at Donnington exchange sidings on 9 July with a single VGA van which it has just brought from Coton Hill yard, Shrewsbury. On this occasion there was no outward traffic from Donnington, and the engine returned light to Shrewsbury. *Paul Shannon*

Shrewsbury: Timber traffic to the rail-connected Kronospan works at Chirk was initially conveyed via Warrington, but in July 1990 the feeder service was altered to run from Shrewsbury instead, connecting out of the early morning Bescot to Shrewsbury train. The new arrangements made good sense because most of Kronospan's timber comes from East Anglia, not Scotland. On the first day of the revised arrangements, 9 July, Distribution-liveried Class 47 No 47238 *Bescot Yard* stands at Coton Hill, Shrewsbury. *Paul Shannon*

Taunton-Bescot: **One of the blue-liveried Class 47s to have acquired an unofficial name, No 47003, *Wild Swan,* passes Wednesbury on 30 July with 6M29, the 14.00 Taunton to Bescot Speedlink feeder service. The traffic on this occasion comprises five empty caustic soda tanks from British Cellophane at Bridgwater, which would have been tripped to Taunton around midday. Often this train carried substantial quantities of cider from Taunton as well. *Paul Shannon***

Crawley-Willesden: **With the extension to Earls Court exhibition centre taking shape on the skyline, Class 47 No 47050 passes West Brompton with 6M88, the 14.42 Crawley to Willesden Speedlink feeder service. On 31 July the traffic is just one empty ferry van, on its way back to mainland Europe via the Dover to Dunkerque train ferry. Flows of timber from Crawley and sand from Holmethorpe once used to swell the load of this trip working, but both had ceased by the date of this photograph. *Paul Shannon***

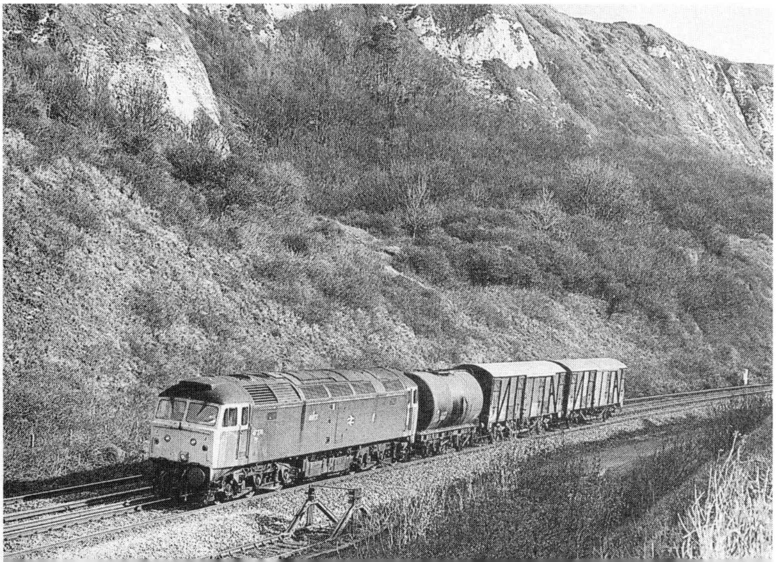

Dover-Paddock Wood: **Class 47 No 47270, unofficially named *Swift,* passes Folkestone Warren with 6W34, the 08.48 Dover Town to Paddock Wood Speedlink trip service. *Brian Denton***

FREIGHTLINER

Multifret 'Swapbodies': **It was intended to introduce Britain's first 90 mph freight service during 1990, carrying 'swapbody' containers on SNCF Multifret wagons. A daily service from Harwich to the North West and Glasgow was envisaged, with high speed enabling a 24-hour turn-round. When the service was eventually introduced in July, the Multifret wagons were limited to 75 mph, in the short term at least, and the train worked only from Harwich to Garston and back. The northbound service, 4M91, the 09.45 from Harwich to Garston, is whisked past South Kenton on 2 November by Class 90 No 90042. The train is made up of ten Multifret wagons which have the capacity to carry 8ft 6in containers in Britain because of their lowered main deck. In this view only conventional 8-foot containers are being carried.** ***Paul Shannon***

Harwich: **Another view of the Multifret train from Harwich to Garston shows the train in Harwich Parkeston Quay yard on 10 August. Class 90 No 90014 awaits departure with the train which has only one container on board on this occasion. The service has acted as a testing ground for the new wagons which are manufactured by Arbel Fauvert Rail of France. At the end of 1990 this company won the contract to produce 700 new Freightliner vehicles.** ***Michael Collins***

Holyhead-Cardiff (1): **During 1990 a new Freightliner service was introduced from Holyhead to Cardiff with a return to working from Cardiff to Crewe Basford Hall yard. On 2 April Class 47 No 47312 is seen shunting at Pengam after arrival with 4V68, the 03.50 Holyhead to Pengam train. It will run round the wagons behind it which form 4M23, the 14.20 Pengam to Basford Hall working, which on this occasion carries mainly steel.** ***Michael Rhodes***

Holyhead-Cardiff (2): **The newly introduced 4V68 Holyhead to Pengam (Cardiff) service (see above) is seen speeding past Dorrington on 15 September behind Class 47 No 47298.** ***Don Gatehouse***

Crewe-Leeds: Class 31 No 31206 approaches Diggle on 9 June with 4E51, the 08.05 MX Crewe to Leeds Freightliner. This service was introduced in January 1990, together with its return working, 4M40, the 15.45 SX Leeds to Crewe. Both services are unusual in being routed via Diggle rather than the Calder Valley line. Basford Hall sidings at Crewe are increasingly being used as an exchange point for Freightliner traffic between ports such as Felixstowe and Southampton, and container terminals in Northern England and Scotland. *Paul Shannon*

Holyhead-Stratford: Freightliner competes for large tonnages of 'Landbridge' traffic between the rest of the world and Ireland. Freightliner services to Holyhead connect with the Dublin and Dun Laoghaire ferries allowing easy transhipment of containerised goods bound for the Emerald Isle. On 15 February Class 85 No 85007 speeds past Wolverton with a heavily loaded 4L86, the 06.30 Holyhead to Stratford service. 1990 was the last year of service for the Class 85/0s. *Mike Goodfield*

Felixstowe-Coatbridge: Class 85/0 No 85037 hauls 4S86, the 10.08 Felixstowe South to Coatbridge Freightliner working near Golborne on 25 May. This daytime service from East Anglia's principal deep-sea port to Scotland's busiest Freightliner terminal was incorporated into the timetable on 15 May, as was a southbound daytime balancing service, 4L58, the 05.23 Coatbridge to Felixstowe. No 85037 was withdrawn from service on 3 September. *Paul Shannon*

Coatbridge-Southampton: Fleet dedication for AC electric locomotives was due to be implemented from the October timetable change, with Railfreight Distribution taking responsibility for sub-Classes 85/1 and 86/6 plus a specified batch of Class 90s. On 8 October, the first Saturday of the new timetable, Class 86 No 86611 *Airey Neave* passes Dallam sidings, Warrington, with 4O81, the 07.10 Coatbridge to Southampton Freightliner service. *Paul Shannon*

Stratford-Felixstowe: During 1990 Class 90 locomotives took over a number of Freightliner diagrams on the Great Eastern main line, displacing the paired Class 86/4s which had been in charge since May 1989. On 30 October Class 90 No 90045, in correct Railfreight Distribution livery, passes Chadwell Heath with 4Y70, the 10.13 Stratford to Felixstowe service. *Paul Shannon*

DEPARTMENTAL INTERLUDE

West Ruislip-Crewe: Distribution-liveried Class 47 No 47225 heads 9X03, West Ruislip-Crewe, past Lower Basildon on Saturday 5 May. This working ran on Sundays for most of 1990, normally hauled by a single Class 31 locomotive, but was rescheduled on this occasion because of the Bank Holiday. *Brian Denton*

Langley Green: The sight of a loose-coupled goods train surrounded by semaphore signals would have been rare in 1980, let alone in 1990. In fact, this is a special working of condemned HTV wagons for Handsworth scrapyard, just setting off from Langley Green yard on 3 July. The motive power is Class 37 No 37198. *Steve Turner*

Perth: Perth hump marshalling yard was opened in March 1962 as part of the 1955 Modernisation Plan, and contained six reception roads and 30 sorting sidings. Because of space restrictions these were placed alongside each other, and in this August 1990 view the remains of the reception yard can be seen on the right. Twenty-six of the original sorting sidings remain intact, but despite the sign outside the yard implying Railfreight Distribution ownership, it has handled virtually no revenue-earning traffic since the closure of the Forfar branch in 1981. *Michael Rhodes*

Radyr: By August 1990 the demise of Radyr yard was all but complete. Network Coal services had been transferred to East Usk and most departmental activity moved to Newport. The only remaining daily freight service to use the yard for shunting was the trip to Cathays wagon works. Here Class 08 No 08660 winds out of the yard on 8 August with an assortment of wagons for repair at Cathays. *Michael Rhodes*

East Usk: East Usk had a short career as the main Speedlink yard for South Wales. Speedlink traffic started to use the yard in November 1987 and was withdrawn a little under three years later in July 1990. Since that time the yard has found new life as the operational centre for Network Coal services in South Wales and as the collection point for ballast from Machen quarry. On 8 August Class 08 No 08942 shunts the afternoon arrival at the east end of the yard. In addition to the ballast and coal, East Usk is still a staging point for Metals traffic to and from Uskmouth. *Michael Rhodes*

THE CLASS 60S ARRIVE

Scunthorpe: Class 60 No 60007 *Robert Adam* was used for crew training based at Immingham and Scunthorpe during the first part of 1990. On 21 March it is seen leaving Scunthorpe West yard with a rake of HEA hoppers bound for Doncaster. The sight of a Class 60 leaving Scunthorpe West yard is a rare one since the yard closed to revenue-earning traffic only a few weeks after the date of the photograph. *Michael Rhodes*

Stevenston: **The first six Class 60s to receive full acceptance by Railfreight were allocated to Thornaby depot and employed on block Metals trains in North East England and Scotland. This included the Hunterston-Ravenscraig iron ore traffic, where Class 60 traction became a regular feature from 14 November onwards. No 60017 *Arenig Fawr* is pictured between Saltcoats and Stevenston with the 6D11 Hunterston-Ravenscraig train on 26 November. *Tom Noble***

Bennerley: **The flagship of BR's newest diesel class, No 60001 *Steadfast,* passes Bennerley on 12 June with an unusual load comprising Lab Coach 6 and 15 Redland PGA hopper wagons. The formation was returning to Mountsorrel after being involved in a series of brake tests on the Perth-Inverness main line. These tests involved a single Class 60 hauling a train equivalent in weight to one of Foster Yeoman's super-trains. The wagons were split into several rakes for the journey between England and Perth, and some of them returned to Mountsorrel empty after discharging their load in Scotland on behalf of BR's civil engineers. *Brian Denton***

Wellingborough: One of the first revenue-earning services to see Class 60 haulage, albeit in tandem with a Class 56, was the daily stone train from Mountsorrel to Radlett. This was ideal for trying out Railfreight's new motive power because it is a straightforward out-and-back diagram, running during daylight hours and easily completed within one shift. On 23 April Class 60 No 60009 *Carnedd Dafydd* pilots Class 56 No 56059 through Wellingborough with 6C31, the 08.58 Mountsorrel-Radlett. No 60009 carries Trainload Construction livery and may well find itself hauling regular stone trains out of Mountsorrel in 1991. *Paul Shannon*

Southampton Millbrook: Class 60 No 60011 *Cader Idris* and Class 47 No 47094 pass Millbrook with the 10.44 Fawley-Eastleigh trip working on 28 September, with a trailing load of just under 2,000 tonnes. Class 60s were used regularly on this train in September and October, first piloting a Class 47 and later singly. *Brian Denton*

Sunderland: Returning to Tees Yard from a crew training trip in the Tyneside area on 4 October are Nos 60007 *Robert Adam* and 60004 *Lochnagar*. The vantage point is Villette Road, south of Sunderland. *Ian Carr*

Motherwell: Class 60 No 60005 *Skiddaw* rests at Motherwell depot on 20 June after a morning's work on crew training runs between Mossend and Millerhill via Shotts. Although Motherwell will carry out day-to-day servicing of Class 60s used on local Metals services, the depot is not expected to receive a specific allocation. *Michael Rhodes*

Old Oak Common: Unaccustomed partners outside Old Oak Common depot on 24 January 1990 are Class 60 No 60003 *Christopher Wren* and Class 59 No 59002 *Yeoman Enterprise.* No 60003's presence was in connection with the training of Old Oak Common drivers on this class. *Michael Collins*

Margam: Trainload Metals used Class 60 No 60014 *Alexander Fleming* for familiarising their South Wales staff with the new motive power. The engine is pictured resting at Margam depot on 7 August. *Michael Rhodes*

CLASS 50 PORTFOLIO

Cowley Bridge Junction: Joining the main line at Cowley Bridge Junction on 29 June is No 50032 *Courageous* with 7C53, the 09.15 Meldon-Exeter ballast train. Despite its Network SouthEast livery, No 50032 was owned by the Departmental sector at the time of the photograph; it was withdrawn on 15 October. *Don Gatehouse*

Ashton Meadows: No 50046 *Ajax* shunts Ashton Meadows engineers' sidings, Bristol, in the early morning of 30 April 1990. Only two visits a week are scheduled to this location which lies adjacent to the former Parson Street-Portishead branch. Clifton Suspension Bridge is visible on the horizon. *Richard Davies*

Bristol: **No 50031** ***Hood*** **rounds the curve from Feeder Bridge Junction to Dr Days Bridge Junction with the 17.00 SuO Bristol East Depot-East Usk departmental working on 7 October.** ***Geoff Dowling***

Stonehouse: **Passing the site of Stonehouse Midland Road station on 1 May is Class 50 No 50015** ***Valiant,*** **in charge of 9B19, the 18.52 Tavistock Junction-Gloucester departmental working. The former Stonehouse coal depot, closed in 1988, lies on the extreme left.** ***Richard Davies***

Stowford: No 50046 *Ajax* passes Stowford with a 10.40 Yeovil Junction to Honiton ballast train. After dropping ballast in this section, the train returned empty to Exeter. *Mike Goodfield*